MEMOIRS

of the
American Mathematical Society

Number 456

WITHDRAWN

Combinatorial Patterns
for Maps of the Interval

Michał Misiurewicz
Zbigniew Nitecki

November 1991 • Volume 94 • Number 456 (second of 4 numbers) • ISSN 0065-9266

American Mathematical Society
Providence, Rhode Island

1980 *Mathematics Subject Classification* (1985 *Revision*).
Primary 58F20, 54H20, 26A18.

Library of Congress Cataloging-in-Publication Data

Misiurewicz, Michał, 1948–
 Combinatorial patterns for maps of the interval/Michał Misiurewicz, Zbigniew Nitecki.
 p. cm. – (Memoirs of the American Mathematical Society, ISSN 0065-9266; no. 456)
 "November 1991, volume 94 (second of 4 numbers)."
 ISBN 0-8218-2513-5
 1. Mappings (Mathematics) 2. Combinatorial analysis. 3. Cycles, Algebraic. I. Nitecki,
Zbigniew. II. Title. III. Series.
 QA3.A57 no. 456
 [QA360]
 510 s–dc20   91-27263
 [511.3′3]   CIP

Subscriptions and orders for publications of the American Mathematical Society should be addressed to American Mathematical Society, Box 1571, Annex Station, Providence, RI 02901-1571. *All orders must be accompanied by payment.* Other correspondence should be addressed to Box 6248, Providence, RI 02940-6248.

SUBSCRIPTION INFORMATION. The 1991 subscription begins with Number 438 and consists of six mailings, each containing one or more numbers. Subscription prices for 1991 are $270 list, $216 institutional member. A late charge of 10% of the subscription price will be imposed on orders received from nonmembers after January 1 of the subscription year. Subscribers outside the United States and India must pay a postage surcharge of $25; subscribers in India must pay a postage surcharge of $43. Expedited delivery to destinations in North America $30; elsewhere $82. Each number may be ordered separately; *please specify number* when ordering an individual number. For prices and titles of recently released numbers, see the New Publications sections of the NOTICES of the American Mathematical Society.

BACK NUMBER INFORMATION. For back issues see the AMS Catalogue of Publications.

MEMOIRS of the American Mathematical Society (ISSN 0065-9266) is published bimonthly (each volume consisting usually of more than one number) by the American Mathematical Society at 201 Charles Street, Providence, Rhode Island 02904-2213. Second Class postage paid at Providence, Rhode Island 02940-6248. Postmaster: Send address changes to Memoirs of the American Mathematical Society, American Mathematical Society, Box 6248, Providence, RI 02940-6248.

10 9 8 7 6 5 4 3 2 1   95 94 93 92 91

CONTENTS

ABSTRACT

The Sharkovskiĭ-like *forcing* relation studied for cycles by Baldwin and others is considered in the general context of finite invariant sets, modelled by *combinatorial patterns*. This relation fails to be antisymmetric, but canonical representatives of equivalence classes are identified; similarly in some cases no single map adequately models the patterns forced by a given one, but simply-stated criteria are formulated for deciding if θ forces η in any specific case. *Reductions* of patterns are introduced; these are reflected in the structure of the Markov graph, and decompose the set of cycles forced by a pattern. A *combinatorial shadowing theorem*, detailing when knowledge of a long piece of orbit that mimics a pattern can be used to force a given permutation, is proved and then used to show that often the set of permutations forced by a given pattern can be approximated using cycles inside the set. The relation between *positive* and *negative representatives* of a given cycle is elucidated, helping to explain an empirical observation by Baldwin; this structure is also related to the doublings and reductions of the cycle. *Maximal patterns* and *permutations* of a given degree are characterized. Finally, the relation between entropy estimates arising from different patterns is explored. In particular, it is shown that $H(\mathfrak{C}_n) \sim log(2n/\pi)$, where $H(\mathfrak{C}_n)$ is the maximum for n-cycles θ of the minimal topological entropy of a map exhibiting θ.

Key words and phrases. Sharkovskiĭ theorem, periodic orbit, interval maps, cycle, combinatorial pattern, adjusted map, horseshoe pattern, fold type, extension of a pattern, reduction of a pattern, combinatorial shadowing, topological entropy.

ACKNOWLEDGEMENT

The project which this paper presents was begun while the authors were attending the Symposium on Ergodic Theory and Dynamical Systems at Warwick during June and July 1986, and subsequently continued at an Oberwolfach conference in August 1986. Major extensions and revisions were carried out during visits by both authors to Göttingen in July and August 1988; the present version of this paper was written during the fall and winter of 1988-9, and final revisions were carried out during the first author's visit to the Institute for Advanced Study, Princeton. We would like to thank all of these institutions for their hospitality and to acknowledge with particular thanks partial support from the British Science and Engineering Council during our stays in Warwick and from Sonderforschungsbereich 170, "Geometrie und Analysis", during our stays at Göttingen. The second author would also like to acknowledge support from Tufts University through a Summer Faculty Fellowship. We thank Juan Tolosa for extensive comments on a preliminary version of this paper. Finally, the second author would like to thank Bill Schlesinger for extensive TEXnical help in the production of this manuscript.

0. Introduction

The celebrated theorem of Sharkovskiĭ [**Sa**] illustrates the rigid restrictions on the set of periodic orbits of a continuous interval map that are imposed by the ordering of points along a line. While the statement of Sharkovskiĭ's theorem focuses on the least periods of these orbits, its various proofs [**Sa; St; Str; HM; Bu; BGMY**] indicate a rich combinatorial structure controlling the disposition of the orbits themselves. Various features of this structure have been elucidated by many workers, beginning with Sharkovskiĭ (and, in a more limited context, Myrberg [**My**] and Metropolis, Stein and Stein [**MSS**]), although the general problem of understanding this structure was explicitly formulated only recently by Baldwin [**Ba**].

Baldwin's work, as well as other recent papers in this area [**ALM; ALS; Bel-3; BlC; BlH; C; Ca; J; F0-2**] have focused on the information inherent in a single periodic orbit. In the present paper, we broaden the scope of this study to encompass arbitrary finite, invariant sets. This leads us to study a relation we call *forcing* on abstract combinatorial objects we call *combinatorial patterns*.

The combinatorial structure we wish to study can be set up as follows. Given $f : I \to I$ a continuous map of a closed interval to itself and $\mathcal{P}$ a finite f-invariant set (i.e., $f[\mathcal{P}] \subset \mathcal{P}$), label the elements of $\mathcal{P}$

$$p_1 < p_2 < \cdots < p_n.$$

Then the action of f on $\mathcal{P}$ can be codified in the map

$$\theta : \{1, \ldots, n\} \to \{1, \ldots, n\}$$

defined by

$$f(p_i) = p_{\theta(i)} \quad i = 1, \ldots, n.$$

The finiteness of $\mathcal{P}$ insures that it consists of finitely many preperiodic f-orbits; the map θ encodes the combinatorial structure of each orbit and the way these orbits intertwine. To stress the combinatorial role of θ, we refer to any map of $\{1, \ldots, n\}$ to itself as a **combinatorial pattern** on n elements, or a *pattern* for short. The number n is the **degree** of θ. We say that the map f **exhibits** the combinatorial pattern θ on $\mathcal{P}$, and call $\mathcal{P}$ a **representative** of θ in f. A given finite invariant set $\mathcal{P}$ represents a unique combinatorial pattern θ, but a given combinatorial pattern θ may have many representatives in f.

When $\mathcal{P}$ is a single periodic orbit, the combinatorial pattern θ represented by $\mathcal{P}$ is a cyclic permutation. More generally, a bijective combinatorial pattern θ is a (possibly multicyclic) permutation, and a representative of θ is a finite union of periodic orbits.

Received by editor February 1, 1990. Received in revised form November 15, 1990.

Given combinatorial data drawn from a single periodic orbit or, more generally, from a finite family of preperiodic orbits for a map f (encoded, respectively, in a cyclic permutation or combinatorial pattern θ that we know to be exhibited by f), what can be said concerning the combinatorics represented by other periodic orbits of f? This question can be formulated more precisely via the abstract notion of *forcing*.

0.1. DEFINITION. *A combinatorial pattern θ* **forces** *another pattern η (denoted $\theta \Rightarrow \eta$) if every continuous map $f : I \to I$ which exhibits θ also exhibits η.*

Our ultimate concern in studying the forcing relation is to understand what kinds of periodic behavior are implied by given combinatorial data. Initially one naturally concentrates on cycles, trying to understand the set of cycles η forced by a given cycle θ, as most of the studies cited earlier have done.

However, it is clear that in many instances the intertwining of several periodic orbits can yield richer information than is implicit in the orbits themselves, regarded separately. Two examples serve to illustrate this. The first is the permutation $(1\ 3)(2\ 5)(4\ 6)$ (see figure 1^1).

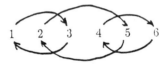

Figure 1

Any 2-cycle alone forces only itself and a fixedpoint; however, it can be shown that the full permutation (consisting of three 2-cycles) forces cycles of all orders. (For example, there exists $x \in (2,3)$ with $f(x) \in (3,4)$, $f^2(x) \in (4,5)$, and $f^3(x) = x$; the existence of cycles of all orders then follows by standard results.) The second example (figure 2) is the pattern θ on 5 elements defined by

$$\theta(1) = 4, \quad \theta(2) = 1 = \theta(5), \quad \theta(3) = 5, \quad \theta(4) = 3.$$

This consists of the "simple" 4-cycle $(1\ 4\ 3\ 5)$ which alone forces only a 2-cycle and a fixedpoint in addition to itself, together with one extra point (2) that maps into this cycle. Again, the existence of this single extra point implies the existence of cycles of all orders.

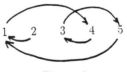

Figure 2

---

[1]Throughout this paper, we use cycle notation for permutations:$(a_1\ a_2\ a_3\ \ldots a_k)$ denotes the cycle $a_1 \mapsto a_2 \mapsto a_3 \mapsto \cdots \mapsto a_k \mapsto a_1$.

These examples illustrate the potential usefulness of admitting combinatorial data about maps in the form of combinatorial patterns. On similar grounds, it might be argued that the combinatorial structure of the set of periodic orbits of a map f is best described by the set of all permutations, not just cycles, exhibited by f, as this encodes not just the relative positions of points within individual orbits, but the placement of different periodic orbits relative to each other.

These philosophical considerations have motivated the formulation of our theory, as far as possible, in the general context of patterns forcing patterns. This broadened perspective has paid practical dividends, clarifying and at times simplifying our work. When stronger, useful results were made possible by specializing to permutations or cycles, we have narrowed our scope. For the most part, however, our study encompasses at least permutations forcing permutations. In any case, it should be pointed out that many of our general results are new even when specialized to cycles.

This paper is in eleven sections. The first two are technical, setting up the machinery we use for studying the forcing relation. It has been customary to use Markov graphs [**BGMY; Bu; HM; N; Str**] or (especially for unimodal maps) the kneading calculus of Milnor and Thurston [**MT; CE; J**] for this purpose. However, these approaches involve technical difficulties, particularly concerning existence and/or uniqueness of itineraries for orbits in $\mathcal{P}$; these difficulties grow as one moves from cycles to permutations and then to patterns. We have adopted a more directly geometric approach, thinking of the objects forced by a given pattern θ as exhibited by a θ-*adjusted map* (see §1), and using partitions generated by preimages of $\mathcal{P}$ in place of loops in the graph or kneading sequences. This approach was initiated in a different context in [**ALM**].

Our main results in §1 are *1.14*, formulating a method for showing that every map exhibiting some given θ also exhibits some η, and *1.19*, a uniqueness result for representatives of θ in a θ-adjusted map. We draw the reader's attention to *1.20*, showing the nonexistence of a "canonical" map for the patterns forced by certain patterns θ, and to *1.21*, which summarizes much of what comes before in a simply stated criterion for deciding, given patterns θ and η, whether θ forces η.

In §2, we study the character of forcing as an order relation. Unlike the case for cycles [**Ba; ALM**], forcing in general is not antisymmetric: $\theta \Rightarrow \eta \Rightarrow \theta$ need not imply $\theta = \eta$ when we do not assume θ and η are cycles. We therefore consider equivalence classes of patterns with respect to forcing. Theorem *2.8* gives two kinds of canonical representatives of each class, and as a corollary *2.9* characterizes the *essential patterns* which are not forced by any proper subpatterns.

In §3, we explore *extensions* (or dually, *reductions*) of a pattern. This idea has been useful in characterizing primary or Sharkovskiĭ-minimal cycles[**ALM; ALS; C; H; ST**]. In a different direction, Bernhardt [**Be1**] has used reductions to determine when a cycle has an immediate successor in the forcing relation. We see that, at least when one is interested in the permutations forced by a given pattern θ, the existence of a reduction of θ can be used to simplify the study of forcing (*3.7*).

In §4, we consider the reductions of a given pattern. Theorem *4.1* relates the non-existence of (nontrivial) reductions of a pattern to strong connectivity of its Markov graph, and Prop. *4.3* shows that (with a short list of exceptions) this is equivalent to primitivity of the transition matrix. In *4.4*, we identify what might be called the *canonical reduction* of a pattern.

In §5 we formulate the notion of *horseshoes* [**MS; M**] as patterns, and the notion of *fold type* of a pattern; we then explore the role of horseshoe patterns in the forcing relation on patterns of fixed fold type. Prop.*5.4* shows the extent to which horseshoes are the forcing-maximal patterns of a given fold type, while Prop.*5.5* gives the restrictions on being able to force a given horseshoe by some permutation of fixed fold type.

In §6, we continue the investigation of extensions and reductions from §4, specializing to the case of cycle reductions. We formulate the notion of a *simple extension* (sometimes called an *orbit extension* [**BlC**] or an *R-extension* [**ALM**]). These play a central role in the characterization of forcing-minimal cycles. We show by examples that in general, a (non-simple) extension of a cycle η by a given pattern or cycle α need not force a simple extension of η by α. However, when α is a horseshoe, such a "simplification" does exist (*6.5*) and it in turn forces simple extensions by all unimodal patterns of a given type (*6.21*).

In §7 we present a new tool, based on a weakening of the notion of extension, which we call *combinatorial shadowing*. The idea here, similar in spirit to [**LMPY**], is that often, to determine that a given map f or pattern θ exhibits (resp. forces) a specific cycle or permutation η, we really only need information about part of the dynamics: for example, we may only need to know about a finite piece of an infinite orbit of f. This is formulated in Theorem *7.10* which says, roughly, that given a pattern η and a permutation π (strongly) forced by η, there is an explicit estimate (depending only on characteristics of π) on how long pieces of orbits of f need to mimic (*shadow*) orbits in η in order to guarantee that f exhibits π. Theorem *7.13* shows that if f has pieces of orbits which shadow a permutation η (without tandem cycles) arbitrarily long then f exhibits η.

In §8, we apply the combinatorial shadowing theorem to study the extent to which the combinatorial data implicit in a pattern θ (specifically, the set of permutations strongly forced by θ) can be approximated "from within" using cycles (strongly) forced by θ. These are formulated in theorems *8.9* and *8.10*, which can be regarded as generalizations of a theorem of Bernhardt for cycles; we explicitly formulate such a generalization in theorem *8.11*. As corollaries, *8.12* and *8.13* show that, for certain permutations θ, the set of cycles (or of permutations) strongly forced by θ characterizes θ uniquely.

In §9 we consider the different ways in which a pattern θ may force a given cycle. We classify (most) periodic orbits of a θ-adjusted map F as positive or negative, and study the relation between *positive* and *negative orbits* that represent the same cycle η in F. The simplest case to state is (Theorem *9.9(3)*) that if η is not a subcycle of θ and η is not a doubling, then the number of positive representatives of η equals the number of negative ones. This sheds light on Baldwin's empirical observation [**Ba**] that when θ is a cycle, most η seem to have an even number of representatives. In general, there is a relation between the representatives of η and the representatives of 2-reductions (and doublings) of η in F. Also (Theorem *9.12*) if η is not a reduction of a subcycle of θ, then associated to any positive representative of η is a simple extension of η by a horseshoe, implying infinitely many (simple) extensions of η also forced by η.

In §10 we consider the forcing relation among patterns, permutations and cycles of a fixed degree. Baldwin [**Ba**] had posed the problem of characterizing *forcing-maximal* cycles of fixed degree; this problem was solved by Jungreis [**J**]. We solve the analogous problems for patterns and for permutations.

Finally in §11 we consider the relation of *topological entropy* to the cycles, permutations and patterns exhibited by a map. We associate to every pattern θ the number $h(\theta)$, which is the minimum topological entropy of maps exhibiting θ. Dually, the entropy of any map f is the supremum of the numbers $h(\theta)$ attached to the patterns it exhibits. We formulate criteria (associated with reductions) which guarantee that $h(\theta)$ is *strictly* greater than $h(\eta)$ for any (appropriate) pattern forced by θ. Finally, we consider a kind of dual to the entropy estimates considered in [**BGMY**] and other papers. Given n, we try to *maximize* (instead of minimizing) $h(\theta)$ among cycles θ of length (i.e. degree) n. While we are not able to explicitly calculate this maximum (some very recent progress in this direction, based on our methods, has been reported [**GT**]), we calculate an asymptotic estimate: this quantity grows with n as $ln(2n/\pi)$.

To close this introduction, we establish some basic terminology and notation. We will deal with continuous maps of a compact interval I to itself, denoting the set of all such maps by $\mathcal{E}(I)$ (these are sometimes called *endomorphisms* of I). The precise description of I is usually unimportant, as all our general results are invariant under conjugacy by scaling; thus we will occasionally not bother to specify I, and speak only of the elements of $\mathcal{E}$.

Given $f \in \mathcal{E}(I)$ and $x \in I$, the **orbit** of x is

$$\mathcal{O}(x) = \{f^j(x) \mid j = 0, 1, \ldots\}.$$

The point x is **periodic** if it is a fixedpoint of some iterate f^k, $k > 0$; it is **preperiodic** if $\mathcal{O}(x)$ is finite. For any preperiodic point x, there exist $j_0 \geq 0$ and $k > 0$ such that

$$f^{k+j}(x) = f^j(x) \quad \text{for all } j \geq j_0.$$

The minimum of all the possible (eventual) periods $k > 0$ is the **least period**, denoted **per**(x), while the minimum of $j_0 \geq 0$ is the time of entrainment, denoted **tail**(x). (This notation comes from thinking of a finite orbit as a kind of limit cycle with a "tail" of transients behind it.) Of course, a periodic point is simply a preperiodic point with $tail(x) = 0$.

We denote by $\mathfrak{P}_n$ (resp. $\mathfrak{S}_n$, $\mathfrak{C}_n$) the set of all patterns (resp., permutations, cycles) of degree n; we will drop the subscript to denote the union over all n. A map f **exhibits** $\theta \in \mathfrak{P}_n$ on $\mathcal{P} = \{p_1 < p_2 < \cdots < p_n\}$ if $f(p_i) = p_{\theta(i)}$ for $i = 1, \ldots, n$; a pattern θ **forces** η $(\theta \Rightarrow \eta)$ if every map exhibiting θ also exhibits η. Given $\theta \in \mathfrak{P}$, the elements of $\mathfrak{P}$ (resp. $\mathfrak{S}$, $\mathfrak{C}$, $\mathfrak{P}_n$, etc.) forced by θ will be denoted $\mathfrak{P}(\theta)$ (resp. $\mathfrak{S}(\theta)$, $\mathfrak{C}(\theta)$, $\mathfrak{P}_n(\theta)$, etc.). Similarly (but very seldom) we denote the patterns (etc.) exhibited by a map f by $\mathfrak{P}(f)$ ($\mathfrak{S}(f)$, etc.). We will call $\eta \in \mathfrak{P}$ (resp. $\mathfrak{S}$, $\mathfrak{C}$) a **subpattern** (resp. **subpermutation, subcycle**) of $\theta \in \mathfrak{P}$ (denoted $\eta \subseteq \theta$) if η is exhibited by the restriction of θ to some subset of $\{1, \ldots, n\}$. A subpattern η of θ is **proper** if $\eta \neq \theta$.

Trivially, θ forces every one of its subpatterns η. We sometimes wish to exclude this trivial version of forcing. If f exhibits θ on $\mathcal{P}$ and η on $\mathcal{Q}$, we can form the union, $\theta \cup \eta$ exhibited by f on $\mathcal{P} \cup \mathcal{Q}$; this depends critically on the relative positions of $\mathcal{P}$ and $\mathcal{Q}$. By a **disjoint union of patterns**, $\theta \vee \eta$, we mean a union as above, with $\mathcal{P} \cap \mathcal{Q} = \emptyset$. We will say that a pattern θ **strongly forces** η (denoted $\theta \Rrightarrow \eta$) if θ forces some disjoint union $\theta \vee \eta$.

Given a set $\mathcal{Q} \subset I$, a map $f : \mathcal{Q} \to \mathcal{Q}$ and $g \in \mathcal{E}(J)$ for some interval J, an **ordered conjugacy** from f to g is a map $h : \mathcal{Q} \to J$ which is strictly increasing ($x < y$ implies $h(x) < h(y)$) into J and satisfies $g \circ h = h \circ f$. We will use this notion most often when $\mathcal{Q}$ is a representative of $\theta \in \mathfrak{P}$ in some $F \in \mathcal{E}(I)$ and $f = F|\mathcal{Q}$; clearly g exhibits θ on $h(\mathcal{Q})$. Conversely, suppose that $F \in \mathcal{E}(I)$ exhibits $\theta \in \mathfrak{P}$ on $\mathcal{Q} \subset I$ and $g \in \mathcal{E}(J)$ exhibits θ on $\mathcal{Q}'$; then clearly there exists a unique ordered conjugacy $h : \mathcal{Q} \to J$ with $h(\mathcal{Q}) = \mathcal{Q}'$.

Several unordered variants of what is usually ordered notation will prove useful. If $a, b \in \mathbb{R}$, define open intervals $\langle a, b \rangle$ by

$$\langle a, b \rangle = \begin{cases} (a, b) & \text{if } a < b \\ (b, a) & \text{if } a > b \\ \emptyset & \text{if } a = b \end{cases}$$

and closed intervals $\operatorname{clos}\langle a, b \rangle$ by

$$\operatorname{clos}\langle a, b \rangle = \begin{cases} [a, b] & \text{if } a < b \\ [b, a] & \text{if } a > b \\ \{a\} & \text{if } a = b \end{cases}$$

Note in particular that $\operatorname{clos}\langle a, a \rangle$ is not just the closure of $\langle a, a \rangle$. Also, we adopt the following convention: if $\sigma = \pm 1$, then

$$x <_\sigma y \quad \text{means} \quad \begin{cases} x < y & \text{if } \sigma = +1 \\ x > y & \text{if } \sigma = -1 \end{cases}.$$

We extend this convention in the obvious way to other inequalities and related notions ($>_\sigma, \leq_\sigma, \sigma$-maximal, etc.)

Finally, we will often pass between a pattern θ and a map f exhibiting θ on some set $\mathcal{P}$. At times, it will be useful to restrict intervals to their intersection with $\mathcal{P}$. Rather than refer to "intervals" in $\mathcal{P}$, we will speak of *blocks*. Thus, given an ordered set $\mathcal{P} = \{x_1 < x_2 < \cdots < x_n\}$ in $\mathbb{R}$, by a **block** in $\mathcal{P}$ we mean a set of adjacent points in $\mathcal{P}$, $B = \{x_i \mid i_0 \leq i \leq i_1\}$.

1. ADJUSTED MAPS

In this section we set up an approach to determining forcing relations based on ideas in [**ALM**] (where only cycles were considered).

Suppose $f \in \mathcal{E}(I)$ exhibits the pattern $\theta \in \mathfrak{P}_n$ on the finite set $\mathcal{P} = \{x_1 < x_2 < \cdots < x_n\}$. By a $\mathcal{P}$-**interval** we mean any of the closed intervals $[x_i, x_{i+1}]$ bounded by adjacent elements of $\mathcal{P}$. The map f is $\mathcal{P}$-**linear** (resp $\mathcal{P}$-**monotone**) if $\mathcal{P}$ includes the endpoints of I and f is affine (resp. monotone) on each $\mathcal{P}$-interval. When we wish to specify the pattern θ exhibited by f on $\mathcal{P}$, we will call f θ-**linear** on $\mathcal{P}$ (resp. θ-**monotone** on $\mathcal{P}$). Note that any two θ-linear maps are topologically conjugate (see [**BlCv1**] for a proof) and in particular they are all conjugate to the **canonical** θ-**linear map** for which $\mathcal{P} = \{1, \ldots, n\}$. (The analogous statement for θ-monotone maps is clearly false.)

Given $f \in \mathcal{E}(I)$, a (nondegenerate) interval $[a_1, a_2] \subset I$ f-**covers** another (nondegenerate) interval $[b_1, b_2]$ with **index** $\sigma = \pm 1$ if $f(a_1) \leq_\sigma b_i \leq_\sigma f(a_2)$ for $i = 1, 2$ (see the introduction for the notation $\leq_\sigma$). A closed interval $\widetilde{I}$ **minimally** f-**covers** the interval J if $\widetilde{I}$ f-covers J and no proper closed subinterval of $\widetilde{I}$ f-covers J. Note that this is equivalent to the conditions that

(i) no interior point of $\widetilde{I}$ maps to the boundary of J and

(ii) $f(\widetilde{I}) = J$.

The following observations will play a central role in our arguments.

1.1. LEMMA. *Suppose $f \in \mathcal{E}$ and $I = [a_1, a_2]$ f-covers $J = [b_1, b_2]$ with index $\sigma = \pm 1$. Then:*

(i) *given any finite subset $y_1 < y_2 < \cdots < y_k$ of J there exists a finite subset $x_1 <_\sigma x_2 <_\sigma \ldots <_\sigma x_k$ of I with $f(x_j) = y_j$ for $j = 1, \ldots, k$;*

(ii) *there exists a subinterval $\widetilde{I} \subset I$ which minimally f-covers J with index σ.*

PROOF:

Proof of (i): We use induction on k. The existence of x_1 follows from the intermediate-value theorem. Suppose we have $x_1 <_\sigma \ldots <_\sigma x_{k-1}$ with $f(x_i) = y_i$. Let α be the endpoint of I with $x_{k-1} <_\sigma \alpha$, and I_k the interval with endpoints x_{k-1}, α. Then I_k f-covers $[y_{k-1}, b_2]$ with index σ, hence contains a pre-image x_k of y_k; since $x_{k-1} <_\sigma x_k \leq_\sigma \alpha$, we are done.

Proof of (ii): Since $f(a_1) \leq_\sigma b_1 \leq_\sigma f(a_2)$, the compact set $f^{-1}[b_1]$ is nonempty; let α be its $<_\sigma$-maximal element. The set $\{x \in I \mid \alpha <_\sigma x$ and $f(x) \geq b_2\}$ is compact and contains an endpoint of I; let β be its σ-minimal element. Then $\alpha <_\sigma \beta$, no element $\alpha <_\sigma x <_\sigma \beta$ can map to either endpoint of J, and α, β map to the endpoints of J. Thus the interval $\widetilde{I}$ with endpoints α, β f-covers J minimally with index σ. ∎

Now, given $f \in \mathcal{E}(I)$, exhibiting $\theta \in \mathfrak{P}_n$ on $\mathcal{P} = \{x_1 < x_2 < \cdots < x_n\}$, we can keep track of f-orbits in $I_\mathcal{P} = [x_1, x_n]$ via itineraries. Abstractly, an **itinerary**

7

of length k is a sequence $\mathcal{I} = \{I_j\}_{j=0}^k$ of $\mathcal{P}$-intervals; its **representative set** for f is

$$\mathcal{R}(f, \mathcal{I}) = \{x \mid f^j(x) \in I_j \text{ for } j = 0, \ldots, k\}.$$

We call the itinerary **proper** if I_j f-covers I_{j+1} for $j = 0, \ldots, k-1$. If $x \in I$ has an f-orbit in $I_\mathcal{P}$ but disjoint from $\mathcal{P}$, there is a unique itinerary of any given length whose representative set contains x.

When f is $\mathcal{P}$-monotone, the expression

$$\mathcal{R}(f, \mathcal{I}) = I_0 \cap f^{-1} \mathcal{R}(f, \{I_j\}_{j=1}^k)$$

combined with induction on k gives the following observations

1.2. REMARK. *If f is $\mathcal{P}$-monotone and $\mathcal{I} = \{I_j\}_{j=0}^k$ is an itinerary of length k, then $\mathcal{R}(f, \mathcal{I})$, if nonempty, is a closed interval whose endpoints map under f^k into $\mathcal{P}$. If $\mathcal{I}$ is proper, then $\mathcal{R}(f, \mathcal{I})$ is a subinterval of I_0 which minimally f^k-covers I_k.*

When f is $\mathcal{P}$-monotone and the f-orbit of x is contained in $I_\mathcal{P} \setminus \mathcal{P}$, then the unique itinerary $\mathcal{I}(x)$ of length k with $x \in \mathcal{R}(f, \mathcal{I}(x))$ is proper. Note in particular that a periodic orbit of f is either contained in $\mathcal{P}$ or disjoint from $\mathcal{P}$. Thus, every periodic orbit of a θ-monotone map ($\theta \in \mathfrak{P}_n$) is identified with either a subcycle of θ or a unique proper itinerary. If $x \notin \mathcal{P}$ is periodic and $per(x) = k$, then $\mathcal{I}(x)$ must have $I_k = I_0$; we call such an itinerary a **loop**. A loop can be traversed many times, creating new loops whose length is a multiple of the basic one; these new loops do not in general determine any new orbits. A loop which is *not* a multiple traversal of some shorter loop is a **prime loop**. A special case is a **primitive loop**, whose elements $I_j, j = 0, \ldots, k-1$ are all distinct.

We would like to see the extent to which itineraries separate periodic points of a θ-linear map.

1.3. LEMMA. *Suppose f is $\mathcal{P}$-linear and $\mathcal{I} = \{I_j\}_{j=0}^k$ is a loop ($I_k = I_0$) whose representative set contains two distinct fixedpoints of f^k. Then $\mathcal{R}(f, \mathcal{I}) = I_0$ and every point of I_j, $j = 0, \ldots, k-1$ is fixed under f^k.*

PROOF: To simplify matters, we conjugate f so that $\mathcal{P} = \{1, \ldots, n\}$. Then the slope of f on each $\mathcal{P}$-interval I_j is an integer λ_j, and the slope of f^k on $\mathcal{R}(f, \mathcal{I})$ is $\lambda = \lambda_0 \cdots \lambda_{k-1}$. Let $x \neq y$ be fixedpoints of f^k in $\mathcal{R}(f, \mathcal{I})$. Then

$$x - y = f^k(x) - f^k(y) = \lambda(x - y)$$

so that $\lambda = 1$. Thus $f^k = id$ on $\mathcal{R}(f, \mathcal{I})$, and so by *1.2* the endpoints of $\mathcal{R}(f, \mathcal{I})$ belong to $\mathcal{P}$. But this means $\mathcal{R}(f, \mathcal{I}) = I_0$. Furthermore, $\lambda = 1$ implies $\lambda_j = \pm 1$ for all j, so $f^j[I_0]$ is a subinterval of I_j of length 1, which is to say $I_j = f^j[I_0]$. This gives $f^k = id$ on I_j for $j = 0, \ldots, k-1$. ∎

The existence of the phenomenon described in *1.3* for any θ-linear map can be detected purely from the combinatorics of θ. Note that by *1.3* if a k-loop is the itinerary of more than one fixedpoint for an iterate f^k of the θ-linear map f, then it consists of intervals whose endpoints belong to cycles in θ of length dividing

k; furthermore, since $f[I_j] = I_{j+1}$, corresponding elements of these cycles are always neighbors in $\mathcal{P}$. We will say $i, i+1 \in \{1, \ldots, n\}$ belong to **tandem cycles** in $\theta \in \mathfrak{P}_n$ if $\theta^j(i+1) = \theta^j(i) \pm 1$ for all j, and both $i, i+1$ are cyclic under θ. Thus, *1.3* says that there exists a k-loop $\mathcal{I} = \{I_j\}_{j=0}^k (I_0 = I_k)$ for which $\mathcal{R}(f, \mathcal{I})$ has two distinct fixedpoints of f^k if and only if the endpoints of I_0 belong to tandem cycles in θ.

If $i, i+1$ are tandem and belong to the *same* cycle of θ, we call this a **self-tandem cycle** in θ; note that if $i, i+1$ belong to tandem cycles and $\theta^\ell(i) = i+1$, then $\theta^\ell(i+1)$ must equal i (otherwise, $\theta^{m\ell}(i) = i + m$ for all m). Thus a self-tandem cycle has least period $k = 2\ell$, and the $\mathcal{P}$-interval bounded by $i, i+1$ is mapped to itself by f^ℓ. We will refer to a $\mathcal{P}$-interval whose endpoints belong to a single self-tandem cycle as a **self-tandem interval**. Since f^ℓ is a linear map interchanging the endpoints of $[i, i+1]$, we see that $f^\ell|[i, i+1]$ is conjugate to the map $g : [-1, 1] \to [-1, 1]$ defined by $g(t) = -t$; we write

$$f^\ell = -id \ \text{ on } [i, i+1]$$

to denote this. In this case, the midpoint has least period ℓ while every other point of $[i, i+1]$ has least period 2ℓ.

On the other hand, if $i, i+1$ belong to *distinct* tandem cycles, and the shorter of these has length ℓ, then I_j are disjoint for $j = 0, \ldots, \ell - 1$, while I_ℓ and I_0 have a common endpoint. Thus either $I_\ell = I_0$ and both cycles have least period $\ell = k$ or I_ℓ and I_0 abut at their common endpoint, and then $I_{2\ell} = I_0$, and the other cycle has length $k = 2\ell$. Thus, when $i, i+1$ belong to distinct tandem cycles then setting k to be the maximum of their lengths, we have every interior point of I_0 periodic with least period k, and its orbit intersects I_0 only once. It is fairly easy to see that in this case each periodic orbit intersecting the interior of I_0 represents the k-subcycle of θ also represented by one of the endpoints of I_0.

From the point of view of describing cycles exhibited by f, self-tandem cycles in θ complicate matters. We would like, therefore, to be able to replace θ with a pattern $\theta_{\frac{1}{2}}$ that has no self-tandem cycles and is equivalent to θ in the sense that $\theta \Rightarrow \theta_{\frac{1}{2}} \Rightarrow \theta$. This is accomplished by the following construction.

1.4. LEMMA. *Suppose f is $\mathcal{P}$-linear, exhibiting $\theta \in \mathfrak{P}_n$ on $\mathcal{P} = \{x_1, \ldots, x_n\}$. Let $\mathcal{P}_{\frac{1}{2}} \supset \mathcal{P}$ be the union of $\mathcal{P}$ with the midpoints of all self-tandem $\mathcal{P}$-intervals. Then:*

(i) *f exhibits a pattern $\theta_{\frac{1}{2}}$ on $\mathcal{P}_{\frac{1}{2}}$ which is independent of the choice of the θ-linear map f, and $\theta_{\frac{1}{2}}$ has no self-tandem cycles;*

(ii) *if $g : I \to I$ is continuous and exhibits θ on a set $\widetilde{\mathcal{P}} = \{y_1 < y_2 < \cdots < y_n\}$, the ordered conjugacy $h : \mathcal{P} \to \widetilde{\mathcal{P}}$ defined by $h(x_i) = y_i$, $i = 1, \ldots, n$ extends to an ordered conjugacy of $\mathcal{P}_{\frac{1}{2}}$ into I;*

(iii) *in particular, $\theta \Rightarrow \theta_{\frac{1}{2}} \Rightarrow \theta$, and $\theta = \theta_{\frac{1}{2}}$ if θ has no self-tandem cycles.*

PROOF:

Proof of (i): The proof of (i) is immediate from our discussion: suppose $\mathcal{P} = \{1, \ldots, n\}$; then $\mathcal{P}_{\frac{1}{2}} \setminus \mathcal{P}$ consists of points $i + \frac{1}{2}$ such that $i, i+1$ belong to a self-tandem cycle. $\mathcal{P}$-linearity of f insures that if $\theta^j(i+1) = \theta^j(i) \pm 1$ then $\theta^j(i+\frac{1}{2}) = \theta^j(i) \pm \frac{1}{2}$, so that $\mathcal{P}_{\frac{1}{2}}$ is invariant, and $\theta_{\frac{1}{2}}$ is clearly independent of our actual scaling of $\mathcal{P}$.

Proof of (ii): Suppose g exhibits θ on $\widetilde{\mathcal{P}}$; by scaling, we can assume $\mathcal{P} = \widetilde{\mathcal{P}} = \{1, \ldots, n\}$, so that $g \in \mathcal{E}(I)$, $\mathcal{P} \subset I$, $g|\mathcal{P} = f|\mathcal{P}$, and $h : \mathcal{P} \to \mathcal{P}$ is the identity map. We need to produce, for every $i + \frac{1}{2} \in \mathcal{P}_{\frac{1}{2}}$, a point $z_i \in (i, i+1)$ with $g^j(z_i)$ between $g^j(i) = \theta^j(i)$ and $g^j(i+1) = \theta^j(i+1) = g^j(i) \pm 1$ such that $g^k(z_i) = z_i$. We do this separately for each cycle in $\mathcal{P}_{\frac{1}{2}} \setminus \mathcal{P}$. Let $\mathcal{I} = \{I_j\}_{j=0}^k$, $I_k = I_0$, be a proper loop bounded by a self-tandem cycle of θ. Since $g = f$ on the endpoints of each I_j, we have that I_j g-covers I_{j+1} with the same index as for the f-covering of I_{j+1} by I_j. Thus, we can use *1.1(ii)* and induction on $j = 0, \ldots, k$ to define intervals $\widetilde{I}_{k-j} \subset I_k$ such that $\widetilde{I}_k = I_k(= I_0)$ and $\widetilde{I}_{k-j-1}$ minimally g-covers $\widetilde{I}_{k-j}$. In particular, $g[\widetilde{I}_{k-j-1}] = \widetilde{I}_{k-j}$. It follows that $\widetilde{I}_k \subset I_0$ g^k-covers $\widetilde{I}_k = I_k = I_0 \supset \widetilde{I}_0$, so $\widetilde{I}_0$ contains a fixedpoint z of g^k, which must satisfy $g^j(z) \in \widetilde{I}_j \subset I_j$. The points $g^j(z)$ clearly fill the role required of $f(\theta^j(i+\frac{1}{2}))$, proving *(ii)*. We shall call these points the **centers** of the self-tandem intervals I_j.

Proof of (iii): This is immediate: $\theta \Rightarrow \theta_{\frac{1}{2}}$ by *(ii)*, and $\theta_{\frac{1}{2}} \Rightarrow \theta$ since $\theta \subseteq \theta_{\frac{1}{2}}$. Furthermore, the construction gives $\mathcal{P}_{\frac{1}{2}} = \mathcal{P}$ if θ has no self-tandem cycles. ∎

Now, we have seen if f is θ-linear on $\mathcal{P}$ and θ has no self-tandem cycles, then any two periodic orbits of f outside $\mathcal{P}$ which have the same itinerary represent the same cycle (which is, in fact, a subcycle of θ). However, a θ-linear map can be modified on any interval bounded by tandem cycles in θ to obtain a θ-monotone map with special properties.

1.5. DEFINITION. *A $\mathcal{P}$-monotone map F is $\mathcal{P}$-adjusted (or θ-adjusted on $\mathcal{P}$) provided*

   (i) *if $\theta(i) \neq \theta(i+1)$ then F is strictly monotone on the $\mathcal{P}$-interval $[x_i, x_{i+1}]$;*
   (ii) *if $x \notin \mathcal{P}$ is F-periodic with least period k, let $\mathcal{I}(x) = \{I_j\}_{j=0}^k$ be the unique (proper) k-loop with $x \in \mathcal{R}(F, \mathcal{I}(x))$. Then $\mathcal{R}(F, \mathcal{I}(x))$ contains no other fixedpoint of F^k.*

1.6. REMARK. *Suppose F is θ-adjusted on $\mathcal{P} = \{x_1 < x_2 < \cdots < x_n\}$. Then:*

   (i) *if $\theta(i) = \theta(i+1)$, then F maps the $\mathcal{P}$-interval $[x_i, x_{i+1}]$ to a point of $\mathcal{P}$, and vice versa:*
   (ii) *if $\mathcal{P}'$ is a finite F-invariant set containing $\mathcal{P}$, then F is automatically $\mathcal{P}'$-adjusted.*

We shall refer to any $\mathcal{P}$-interval that is mapped to a point by F as a **flat $\mathcal{P}$-interval** (the graph of F over a flat $\mathcal{P}$-interval is a horizontal line segment), and will use the notation

$$Flat(F, \mathcal{P}) = \{x \mid x \text{ is interior to a flat } \mathcal{P}\text{-interval}\}.$$

Note that if F, F' are both $\mathcal{P}$-adjusted and agree on $\mathcal{P}$ then $Flat(F, \mathcal{P}) = Flat(F', \mathcal{P})$.

The existence of $\mathcal{P}$-adjusted maps is formulated in the following.

1.7. LEMMA. *For any $\mathcal{P}$-linear map f, there exists a $\mathcal{P}$-adjusted map F which agrees with f on any $\mathcal{P}$-interval whose endpoints do not belong to tandem cycles.*

The specific θ-adjusted map produced by the construction in the following proof, starting from the $\mathcal{P}$-linear map with $\mathcal{P} = \{1, \ldots, n\}$, will be called the **canonical θ-adjusted map.**

PROOF: Note first that if F is $\mathcal{P}_{\frac{1}{2}}$-adjusted, where $\mathcal{P}_{\frac{1}{2}}$ is as in *1.4*, then it is automatically $\mathcal{P}$-adjusted; so we work with $\mathcal{P}_{\frac{1}{2}}$ in place of $\mathcal{P}$. Also up to conjugacy, we can work with $\mathcal{P}_{\frac{1}{2}} = \{1, \ldots, n\}$. Construct F as follows:

(i) On any $\mathcal{P}_{\frac{1}{2}}$-interval $J = [i, i+1]$, $f = F$ unless $i, i+1$ belong to tandem cycles in $\theta_{\frac{1}{2}}$ and $i \leq \theta^j(i), \theta^j(i+1)$ for all j.

(ii) If $i, i+1$ belong to tandem cycles in $\theta_{\frac{1}{2}}$ and i is the leftmost point in the union of these cycles, then the (least) period k of i is at least equal to the (least) period of $i+1$; since $\theta_{\frac{1}{2}}$ has no self-tandem cycles, we have $f^j(\text{int } J)$ disjoint for $j = 0, \ldots, k-1$, and $f^k = id$ on J. For $0 \leq t \leq 1$, let $F(i+t) = f(i+t^2)$.

Then we see that on any $\mathcal{P}_{\frac{1}{2}}$-interval $J = [i, i+1]$ bounded by tandem cycles, $F^k(i+t) = i+t^2$. Hence, no periodic orbit of F intersects $\text{int } J$. Using *1.3*, this gives us that F is $\mathcal{P}$-adjusted. ∎

The basic property of $\mathcal{P}$-adjusted maps is that periodic orbits outside $\mathcal{P}$ are distinguished by their itineraries. This is made precise in the following.

1.8. LEMMA. *Suppose F is $\mathcal{P}$-adjusted and $x \notin \mathcal{P}$ is a fixedpoint of F^k. Let $\mathcal{I} = \{I_j\}_{j=0}^k (I_0 = I_k)$ be the (unique) k-loop with $x \in \mathcal{R}(F, \mathcal{I})$. Then $\mathcal{R}(F, \mathcal{I})$ contains no other fixedpoint of F^k, unless the endpoints of $\mathcal{R}(F, \mathcal{I})$ belong to a self-tandem cycle in $\mathcal{P}$ and k is a multiple of $2 \cdot per(x)$. In this exceptional case, the two endpoints of $\mathcal{R}(F, \mathcal{I})$ together with x are the only fixedpoints of F^k in $\mathcal{R}(F, \mathcal{I})$.*

PROOF: Set $\ell = per(x)$ and let $\mathcal{I}_0 = \{I_0, \ldots, I_\ell = I_0\}$ be the ℓ-loop with $x \in \mathcal{R}(F, \mathcal{I}_0)$. Note that ℓ divides k and $\mathcal{I}$ is a repetition of $\mathcal{I}_0$: $I_{j+\ell} = I_j$ for $j \leq k - \ell$. We claim that $\mathcal{I}_0$ is prime: if $\mathcal{I}_0$ were a repetition of a shorter loop $\widetilde{\mathcal{I}}$, then $\mathcal{R}(F, \widetilde{\mathcal{I}})$ would contain a periodic point y of F with $per(y)$ dividing ℓ, and $y \neq x$ because $per(y) < per(x)$; but then $F^\ell(y) = y$, $F^\ell(x) = x$, and both would belong to $\mathcal{R}(F, \mathcal{I}_0)$, contrary to *1.5*.

This shows that ℓ is the least period of $\mathcal{I}$, and so ℓ divides $per(y)$ for any fixedpoint y of F^k in $\mathcal{R}(F, \mathcal{I})$. Suppose $y \neq x$ is such a fixedpoint, with $per(y) = m\ell$. We have then that $m > 1$ and $x \neq y$ are fixedpoints of $F^{m\ell}$ in $\mathcal{R}(F, \mathcal{I}(y))$, where $\mathcal{I}(y)$ is the m-fold repetition of $\mathcal{I}_0$. If $y \notin \mathcal{P}$, we have a contradiction to *1.5*, so $y \in \mathcal{P}$. But then $y_i = f^{i\ell}(y) \in I_{i\ell} \cap \mathcal{P}$ for $0 \leq i \leq m$, and $y_0, \ldots, y_{m-1}$ are distinct. Since I_0 is a $\mathcal{P}$-interval y_i must be an endpoint of I_0 for each i; thus $m > 1$ forces $m = 2$. But then $\mathcal{R}(F, \mathcal{I}) = I_0$; furthermore, for all j, $F^j I_0$ is a subinterval of I_j containing the distinct points $F^j(y_0), F^j(y_1) \in \mathcal{P}$. Thus, y_0 and y_1 belong to a tandem cycle in P. ∎

We now would like to use itineraries to distinguish points under a $\mathcal{P}$-adjusted map F and to simultaneously locate combinatorially analogous points in any map g exhibiting the same pattern. To this end, we construct a sequence of invariant sets $\mathcal{P}_k$ which distinguish k-itineraries.

1.9. DEFINITION. *Suppose F is θ-adjusted on $\mathcal{P} = \mathcal{P}_0$. Define an (increasing) sequence of finite F-invariant sets $\mathcal{P}_k$, $k = 1, 2, \ldots$ inductively by*

$$\mathcal{P}_{j+1} = \{x \mid x \notin Flat(F, \mathcal{P}_j) \text{ and } F(x) \in \mathcal{P}_j\}, \quad j = 0, 1, \ldots.$$

The pattern exhibited by F on $\mathcal{P}_k$ is denoted θ_k.

Note that $Flat(F, \mathcal{P}_k) = Flat(F, \mathcal{P})$ for all k. Furthermore, F^k is $\mathcal{P}_k$-adjusted.

The following result, analogous to [**ALM**,1.18], tells us that this construction is " universal" for forcing. Note that if F is θ-adjusted on $\mathcal{P}$ and g exhibits θ on $\mathcal{Q}$, then there is a unique ordered conjugacy $h : \mathcal{P} \to \mathcal{Q}$.

1.10. LEMMA. *Suppose $\theta \in \mathfrak{P}_n$, f is θ-adjusted on $\mathcal{P} = \{x_1 < x_2 < \cdots < x_n\}$ and $g \in \mathcal{E}(I)$ exhibits θ on $\widetilde{\mathcal{P}} = \{y_1 < y_2 < \cdots < y_n\}$. Then, for each $k = 1, 2, \ldots$, the unique ordered conjugacy $h : \mathcal{P} \to \widetilde{\mathcal{P}}$ defined by $h(x_i) = y_i$, $i = 1, \ldots, n$ extends to an ordered conjugacy $h_k : \mathcal{P}_k \to I$. Thus, $h_k(\mathcal{P}_k) = \widetilde{\mathcal{P}}_k$ is a representative of θ_k in g. In particular, if g is also θ-adjusted, h_k can be chosen so that $\widetilde{\mathcal{P}}_k = (\widetilde{\mathcal{P}})_k$–that is, the construction in 1.9 yields the same pattern θ_k for all θ-adjusted maps.*

PROOF: First, note that it suffices to prove the lemma for $k = 1$, since then replacing θ with θ_{k-1} and $\mathcal{P}$ with $\mathcal{P}_{k-1}$ we get the necessary induction on k.

For each non-flat $\mathcal{P}$-interval, $J = F[x_i, x_{i+1}] = [x_{i'}, x_{i'+s}]$ for some $s \geq 1$, and J is F-covered by $I = [x_i, x_{i+1}]$ with index $\sigma = \pm 1$; the corresponding interval $\widetilde{J} = [y_{i'}, y_{i'+s}]$ is g-covered by $\widetilde{I} = [y_i, y_{i+1}]$, also with index σ. Apply *1.1(i)* to f, I, J and $g, \widetilde{I}, \widetilde{J}$ to obtain a σ-ordered set of preimages in I(resp. $\widetilde{I}$) of $\mathcal{P} \cap J$(resp. $\widetilde{\mathcal{P}} \cap \widetilde{J}$). In the first case, this is precisely $\mathcal{P}_1 \cap I$ since $F|I$ is one-to-one, and in the second it gives the required points for $\widetilde{\mathcal{P}}_1 \cap \widetilde{I}$. The ordering of these two sets gives the unique extension of h to $h_1 : \mathcal{P}_1 \to \widetilde{\mathcal{P}}_1$.

If g is also θ-adjusted then the choice of preimages is unique, and hence $\widetilde{\mathcal{P}}_1 = (\widetilde{\mathcal{P}})_1$. ∎

We will use *1.10* in showing that θ forces certain patterns exhibited by a θ-adjusted map F. To this end, we first establish three technical results that clarify the relation between the $\mathcal{P}$- and $\mathcal{P}_k$-itineraries of various orbits.

1.11. LEMMA. *Suppose F is θ-adjusted on $\mathcal{P}$ and x is a periodic point for F with $\mathcal{O}(x) \cap \mathcal{P} = \emptyset$ and $per(x) = k$. Let $\mathcal{I} = \{I_i\}_{i=0}^{2k}$ be the (unique, proper) $\mathcal{P}$-itinerary of x, and for $i = 0, \ldots, 2k$ let J_i be the $\mathcal{P}_{2k-i}$-interval containing x. Then:*

(i) $J_i = F^i(J_0) \subset I_i$ for $i = 0, \ldots, 2k-1$, and $F^{2k}(J_0) = J_{2k} = I_{2k} = I_0$;

(ii) $J_0 \subset \text{int } I_0$, unless $I_0 = J_0$ has endpoints belonging to a self-tandem cycle of length $2k$.

PROOF: Note first that since x is periodic, $\mathcal{O}(x) \cap Flat(F, \mathcal{P}) = \emptyset$.

Now, it is clear that $J_{2k} = I_{2k} = I_0$ and that F maps J_i onto J_{i+1} in a strictly monotone way; *(i)* follows immediately.

If the endpoints of I_0 belong to a self-tandem cycle then clearly $J_0 = I_0$; so to prove *(ii)* it remains to show that if the endpoints of I_0 do not belong to a self-tandem cycle of length $2k$ then J_0 is interior to I_0. But we know already that F^{2k} is strictly monotone on J_0, and x is a fixedpoint of F^k interior to J_0. Thus, F^{2k} is strictly increasing near x, hence on all of J_0. It follows that any endpoint of J_0 belonging to $\mathcal{P}$ is a fixedpoint of F^{2k}, contradicting *1.8*. ∎

1.12. LEMMA. *Suppose F is $\mathcal{P}$-adjusted and $a \neq b$ satisfy*

$$F^i(a), F^i(b) \notin \mathcal{P} \cup Flat(F, \mathcal{P}) \ \text{ for } i = 0, \dots, n-1$$

(where $n \geq 1$). Then for each $k \geq 0$,

$$card \ \{\mathcal{P}_{k+n} \cap \langle a, b \rangle\} \geq \ card \ \{\mathcal{P}_k \cap \langle F^n(a), F^n(b) \rangle\}.$$

Here, *card S* denotes the cardinality of the set S.

PROOF: We prove the statement for $n = 1$; the case for general $n \geq 1$ follows by induction.

Pick $k \geq 0$. For each $x \in \mathcal{P}_k \cap \langle F(a), F(b) \rangle$, x has a pre-image y in $\langle a, b \rangle$:

$$y \in F^{-1}(x) \cap \langle a, b \rangle.$$

If $y \notin Flat(F, \mathcal{P})$, then $\alpha(x) = y$ satisfies

$$\alpha(x) \in F^{-1}(x) \cap \langle a, b \rangle \cap \mathcal{P}_{k+1}.$$

On the other hand, if $y \in Flat(F, \mathcal{P})$, then since $a, b \notin Flat(F, \mathcal{P})$, there exists $z \in \langle a, b \rangle \cap \mathcal{P}_{k+1}$ with $f(z) = f(y) = x$. Thus $\alpha(x) = z$ again satisfies

$$\alpha(x) \in F^{-1}(x) \cap \langle a, b \rangle \cap \mathcal{P}_{k+1}.$$

This process picks out a distinct element $\alpha(x)$ of $\mathcal{P}_{k+1} \cap \langle a, b \rangle$ for every $x \in \mathcal{P}_k \cap \langle a, b \rangle$, and shows the desired inequality. ∎

The following result allows us to control the separation of distinct preperiodic points by the refined sets $\mathcal{P}_k$.

1.13. PROPOSITION. *Suppose F is $\mathcal{P}$-adjusted and $a \neq b$ are both preperiodic with orbits disjoint from $\mathcal{P} \cup Flat(F, \mathcal{P})$. Let*

  $n = $ *least common multiple of $per(a)$, $per(b)$*
  $m = max\{tail(a), \ tail(b)\}$.

Then

  (i) $\mathcal{P}_N \cap \langle a, b \rangle \neq \emptyset$, where $N = n + m - 1$;
  (ii) if $m = 0$ (i.e., a and b are both periodic), then card $\{\mathcal{P}_M \cap \langle a, b \rangle\} \geq 2$, where $M = 2n - 1$.

PROOF:

Proof of (i): We distinguish two cases.

Case 1: $F^k(a) = F^k(b)$ *for some* $k > 0$: Assume k is the least such (positive) integer, so that $F^{k-1}(a) \neq F^{k-1}(b)$. Since F is strictly monotone on $\mathcal{P}$-intervals containing $F^{k-1}(a)$ or $F^{k-1}(b)$, these points must belong to distinct $\mathcal{P}$-intervals, so that

$$\mathcal{P} \cap \langle F^{k-1}(a), F^{k-1}(b) \rangle \neq \emptyset,$$

and hence by *1.12*

$$\mathcal{P}_{k-1} \cap \langle a, b \rangle \neq \emptyset$$

as required.

Case 2: $F^k(a) \neq F^k(b)$ *for all* $k \geq 0$: Let

$$x = F^m(a), \quad y = F^m(b).$$

Then $x \neq y$ are distinct fixedpoints of F^n. Lemma *1.8* tells us that the n-itineraries of x and y are distinct, so for some $\ell < n$,

$$\mathcal{P} \cap \langle F^\ell(x), F^\ell(y) \rangle \neq \emptyset.$$

By *1.12*,

$$\mathcal{P}_{\ell+m} \cap \langle a, b \rangle \neq \emptyset,$$

and since $\ell + m \leq N$, we get

$$\mathcal{P}_N \cap \langle a, b \rangle \neq \emptyset,$$

proving *(i)*.

Proof of (ii): By assumption, we have $F^n(a) = a$, $F^n(b) = b$, and by *(i)*, there exists

$$x \in \mathcal{P}_N \cap \langle a, b \rangle.$$

Now, if $F^n(x) = x$, let $\mathcal{I}$ (resp. $\mathcal{J}$) be the itinerary of a(resp. b) of length n. By *1.8*, if $x \in \mathcal{R}(F, \mathcal{I})$, then $x \in \mathcal{P}$ and its orbit is self-tandem, with $a \in \langle x, F^k(x) \rangle$, where $per(x) = 2k$, and an analogous statement holds if $x \in \mathcal{R}(F, \mathcal{J})$. Since $x \in \langle a, b \rangle$, x cannot simultaneously belong to $\mathcal{R}(F, \mathcal{I})$ and $\mathcal{R}(F, \mathcal{J})$.

Assume, then, that $x \notin \mathcal{R}(F, \mathcal{I})$. Then for some $\ell < n$ we must have

$$\mathcal{P} \cap \langle F^\ell(a), F^\ell(x) \rangle \neq \emptyset$$

and hence by lemma *1.12*

$$\mathcal{P}_\ell \cap \langle a, x \rangle \neq \emptyset.$$

Since $N \leq M$ and $\ell \leq n - 1 \leq M$, we thus get x and one additional point (between a and x) in $\mathcal{P}_M \cap \langle a, b \rangle$, as required. This proves *(ii)* when $F^n(x) = x$.

If $F^n(x) \neq x$, then either $x \in \langle a, F^n(x) \rangle$ or $x \in \langle b, F^n(x) \rangle$; assume the former. Since $F^n(a) = a$, $x \in \langle F^n(a), F^n(x) \rangle$ and lemma *1.12* gives

$$card \{\mathcal{P}_{N+n} \cap \langle a, x \rangle\} \geq \ card \{\mathcal{P}_N \cap \langle F^n(a), F^n(x) \rangle\} \geq 1.$$

Again this gives us a point in addition to x in $\mathcal{P}_{N+n} \cap \langle a, b \rangle$, and since $N + n = M$, this completes the proof of *(ii)*. ∎

A major tool in our analysis is the following.

1.14. THEOREM. *Given $\theta \in \mathfrak{P}$, let F be θ-adjusted on $\mathcal{P}$, and let $\mathcal{Q}$ be any finite F-invariant set disjoint from $Flat(F, \mathcal{P})$.*

For any $g \in \mathcal{E}(I)$ exhibiting θ on $\widetilde{\mathcal{P}}$, the (unique) ordered conjugacy $h : \mathcal{P} \to \widetilde{\mathcal{P}}$ extends to an ordered conjugacy $h : \mathcal{P} \cup \mathcal{Q} \to I$. In particular, θ forces the pattern represented by $\mathcal{P} \cup \mathcal{Q}$ in F.

PROOF: Let $\mathcal{R}$ be the set of periodic points in $\mathcal{Q} \setminus \mathcal{P}$. Then $\mathcal{R}$ is a finite union of periodic orbits and by *1.13* there exists N such that any $\mathcal{P}_N$-interval intersects $\mathcal{R}$ in at most one point.

Suppose $x \in \mathcal{R}$, and let $k = per(x)$. Then there exists a unique (proper) k-loop $\{I_j\}_{j=0}^k$ of $\mathcal{P}_N$-intervals with $F^i(x) \in I_i$. By *1.10* we can find a corresponding k-loop $\{\widetilde{I}_j\}_{j=0}^k$ of $\widetilde{\mathcal{P}}_N$-intervals such that $\widetilde{I}_i$ g-covers $\widetilde{I}_{i+1}$ for $i = 0, \ldots, k - 1$. A standard argument (see [**BGMY**]) gives a point $z \in \widetilde{I}_0$ with $g^i(z) \in \widetilde{I}_j$, $j = 0, \ldots, k - 1$ and $g^k(z) = z$. Now, none of the points $g^j(z)$ can be an endpoint of $\widetilde{I}_j$, since otherwise $z = g^k(z) \in \widetilde{\mathcal{P}}$, which by *1.10* would give $y \in \mathcal{P}$ with $F^j(y) \in I_j$, $j = 0, \ldots, k-1$ and $F^k(y) = y$, in contradiction to *1.8*. Thus, for every periodic F-orbit $\mathcal{O}(x)$ in $\mathcal{R}$, we obtain a corresponding periodic g-orbit $\mathcal{O}(z)$ with a corresponding itinerary.

Carrying out this construction with one point from each periodic orbit in $\mathcal{R}$ gives us a natural extension of h to an ordered conjugacy on $\mathcal{P}_N \cup \mathcal{R}$. Now, by remark *1.6(ii)*, F is $(\mathcal{P}_N \cup \mathcal{R})$-adjusted, and so by *1.10* we can further extend h to an ordered conjugacy on $(\mathcal{P}_N \cup \mathcal{R})_M$ for some M, and the restriction of our ordered conjugacy to $\mathcal{P} \cup \mathcal{Q}$ is the map required by the theorem. ∎

A useful special case of *1.14* is the following.

1.15. COROLLARY. *Suppose $\theta \in \mathfrak{P}$ and $\eta \in \mathfrak{P}_k$ such that $\eta(i) \neq \eta(i+1)$ for $i = 0, \ldots, k-1$. Let F be any θ-adjusted map. Then $\theta \Rightarrow \eta$ if and only if F exhibits η.*

PROOF: This follows from the observation that if a representative $\mathcal{Q}$ of a pattern η in F intersects some flat $\mathcal{P}$-interval in more than one point, then $\eta(i) = \eta(i+1)$ for some i. But if $\mathcal{Q}$ intersects each flat $\mathcal{P}$-interval in at most a single point, we can alter $\mathcal{Q}$, replacing points of $\mathcal{Q}$ interior to flat $\mathcal{P}$-intervals with their endpoints, in $\mathcal{P}$, and adjusting pre-images. Thus, we obtain a representative $\mathcal{Q}'$ of η satisfying the hypotheses of *1.14*. The pattern exhibited by F on $\mathcal{P} \cup \mathcal{Q}'$ is forced by θ and contains η as a subpattern. ∎

We remark also that *1.14* gives us a complete characterization of the patterns forced by a permutation.

1.16. COROLLARY. *If $\theta \in \mathfrak{C}$ and F is θ-adjusted, then $\eta \in \mathfrak{P}$ is forced by θ if and only if η is exhibited by F.*

PROOF: If $\theta \in \mathfrak{C}$ then F has no flat intervals. Thus any representative $\mathcal{Q}$ of η in F satisfies the hypotheses of *1.14*. ∎

To study the patterns forced by a general pattern θ, we must study more carefully the role of flat intervals in the forcing relation. This is the purpose of the next two technical results; they lead up to *1.19* which remains of interest even when θ is a permutation.

1.17. LEMMA. *Given* $\theta \in \mathfrak{P}$ *suppose* F *is* θ-*adjusted on* $\mathcal{P}$. *Then for any integer* k, *the set*

$$\mathcal{R}_k = \{x \mid card\, \mathcal{O}(x) \le k \text{ and } \mathcal{O}(x) \cap Flat(F, \mathcal{P}) \ne \emptyset\}$$

is finite.

PROOF: By lemma *1.8*, the number of periodic points for F with a given period is finite. Since F is strictly monotone on non-flat $\mathcal{P}$-intervals, the number of inverse images outside $Flat(F, \mathcal{P})$ of any point is also finite. Finally, if $card(\mathcal{O}(x)) \le i+1$ then either x is periodic or $card(\mathcal{O}(F(x))) \le i$. Now the lemma follows by an easy induction on k. ∎

We now give a construction which plays a key role in the following three results.

1.18. LEMMA. *Suppose* $\theta \in \mathfrak{P}$ *and* $F \in \mathcal{E}(I)$ *is* θ-*adjusted on* $\mathcal{P}$. *For every positive integer* k, *there exists* $G_k \in \mathcal{E}(I)$ *such that*

(i) $G_k = F$ *outside* $Flat(F, \mathcal{P})$;
(ii) $x \in Flat(F, \mathcal{P})$ *implies* $card\{\mathcal{O}_{G_k}(x)\} > k$.

PROOF: We will construct G_k by first showing that *(ii)* follows from *(i)* together with a finite family of pointwise estimates, then constructing a map satisfying these estimates. Fix k. By *1.17*, for each i the set

$$\mathcal{R}_i = \{x \mid card\{\mathcal{O}_F(x)\} \le i \text{ and } \mathcal{O}_F(x) \cap Flat(F, \mathcal{P}) \ne \emptyset\}$$

is finite. In particular, there exists $\varepsilon > 0$ such that for any $x \in \mathcal{R}_{k-1} \setminus \mathcal{P}$ and $y \in \mathcal{P}$ we have

$$|x - y| \ge \varepsilon.$$

Now suppose $G \in \mathcal{E}(I)$ satisfies *(i)* and the following two conditions:

(iii) $\sup_x |G(x) - F(x)| < \varepsilon$;
(iv) for every $x \in Flat(F, \mathcal{P})$, $G(x) \ne F(x)$.

We claim that under these conditions if $x \notin \mathcal{R}_k$, then either $card\{\mathcal{O}_G(x)\} > k$ or $\mathcal{O}_G(x)$ includes a periodic G-orbit which intersects $Flat(F, \mathcal{P})$ and has period less than k. For suppose $x \notin \mathcal{R}_k$ and $\mathcal{O}_G(x)$ has at most k points; let i be the largest integer satisfying $G^i(x) \in Flat(F, \mathcal{P})$, if it exists. By *(iii)* and *(iv)*, $y = G^{i+1}(x) \notin \mathcal{R}_{k-1}$, and $\mathcal{O}_G(y) = \mathcal{O}_F(y)$ is disjoint from $Flat(F, \mathcal{P})$. It follows that

$$card\{\mathcal{O}_G(x)\} = i + 1 + card\{\mathcal{O}_F(y)\} > i + 1 + k - 1 \ge k,$$

contradicting our assumption on x. Thus, the maximal i described above does not exist. If $\mathcal{O}_G(x) \cap Flat(F, \mathcal{P}) = \emptyset$, then $x \notin \mathcal{R}_k$ means $card\{\mathcal{O}_G(x)\} = card\{\mathcal{O}_F(x)\} > k$, contradicting our choice of k. Thus, $G^i(x) \in Flat(F, \mathcal{P})$ for infinitely many i, and it follows that $\mathcal{O}_G(x)$ contains a periodic orbit of least period at most k intersecting $Flat(F, \mathcal{P})$, as claimed.

Thus, if in addition to *(i)*, *(iii)* and *(iv)* we can insure

(v) every periodic orbit of G with period $\le k$ is disjoint from $Flat(F, \mathcal{P})$

then $\mathcal{R}_k$ includes all points x for which $card(\mathcal{O}_G(x)) \leq k$, so that *(ii)* is satisfied.

We will see that *(v)* is implied by *(i),(iii),(iv)* and

> *(vi)* if J is an open $\mathcal{P}$-interval, flat for F, such that $G^j(J)$ is disjoint from any F-flat $\mathcal{P}$-intervals for $j = 1, \ldots, i-1$ and $G^j(J)$ intersects the F-flat $\mathcal{P}$-interval K, then G^i contracts J into K.

To see that *(vi)* implies *(v)* (assuming *(i),(iii)* and *(iv)*), suppose that $G^m(x) = x$, $m \leq k$, belongs to the F-flat interval J. Applying *(vi)* to every F-flat interval that intersects $\mathcal{O}(x)$, we find that G^m contracts J into J. Now F^m and G^m agree on the endpoints of J and F^m maps these endpoints into $\mathcal{P}$. But x is interior to J and G^m is a contraction; thus the point of $\mathcal{P}$ nearest x (one of the endpoints of J) must map interior to J contrary to F^m-invariance of $\mathcal{P}$. This contradiction shows that *(i),(iii),(iv)* and *(vi)* imply *(v)*, hence *(ii)*.

It remains to construct G satisfying *(i),(iii),(iv)* and *(vi)*.

Let J be a $\mathcal{P}$-interval, flat for F. Thus, $F[J]$ is a single point, $p(J) \in \mathcal{P}$. For $0 < \delta$, let

$$L_\delta(J) = [p(J), p(J) + \delta]$$

unless $p(J)$ is the rightmost point of $\mathcal{P}$, in which case

$$L_\delta(J) = [p(J) - \delta, p(J)].$$

Now, let y be the midpoint of J and 2ζ the length of J so

$$J = [y - \zeta, y + \zeta].$$

We will define G on $J_+ = [y - \zeta, y]$ and extend by symmetry (i.e., $G(y + t) = G(y - t)$ for $0 \leq t \leq \zeta$).

Suppose first that for some $\delta > 0$ sufficiently small,

$$F^i(L_\delta) \cap Flat(F, \mathcal{P}) = \emptyset \text{ for } i = 0, \ldots, k - 1.$$

Then pick $\delta < \varepsilon$ so that this condition holds and map J_+ affinely onto L_δ, letting $G(y - \zeta) = p(J)$.

On the other hand, if for some $i \in \{1, \ldots, k-1\}$, $F^i(L_\delta)$ intersects $Flat(F, \mathcal{P})$ for all $\delta > 0$, pick the least such i, and note that for $\delta > 0$ sufficiently small, F^i maps L_δ homeomorphically onto a subinterval of $Flat(F, \mathcal{P})$ with one endpoint $F^i(p(J))$ in $\mathcal{P}$. Reducing δ, we can assume $0 < \delta < \varepsilon$ and $|F^i(L_\delta)| < \zeta$. Now, we can define G on J_+ so that J_+ maps onto L_δ homeomorphically, with $G(y - \zeta) = p(J)$, so that $F^i{\circ}G$ is affine. The extension by symmetry then has $F^i{\circ}G$ affine on $[y - \zeta, y]$ and $[y, , y + \zeta]$, with Lipschitz constant less than 1. We clearly have *(i), (iii), (iv)* and *(vi)* satisfied, so that G is our required map G_k. ∎

Using *1.18*, we can now establish the following basic observation.

1.19. THEOREM. *For any $\theta \in \mathfrak{P}$ and $F \in \mathcal{E}(I)$, θ-adjusted on $\mathcal{P}$, the set $\mathcal{P}$ is the only representative of θ in F which is disjoint from $Flat(F, \mathcal{P})$.*

PROOF: Suppose $\theta \in \mathfrak{P}_n$, and consider the set $\mathcal{R}_n$ of points whose F-orbit has at most n elements and is disjoint from $Flat(F, \mathcal{P})$. By *1.17*, $\mathcal{R}_n$ is finite, and so $\mathcal{Q} = \mathcal{R}_n$ satisfies the hypotheses of *1.14*.

Let $\widetilde{\mathcal{P}}$ be any representative of θ in F disjoint from $Flat(F,\mathcal{P})$, and note that $\widetilde{\mathcal{P}} \subset \mathcal{R}_n$. Let $G = G_n \in \mathcal{E}(I)$ be the map given by 1.18 with $k = n$. Since $\widetilde{\mathcal{P}} \cap Flat(F,\mathcal{P}) = \emptyset$, $G = F$ on $\widetilde{\mathcal{P}}$, so $\widetilde{\mathcal{P}}$ is a representative of θ in G. Now apply 1.14 to extend the (unique) ordered conjugacy $h : \mathcal{P} \to \widetilde{\mathcal{P}}$ to an ordered conjugacy $h : \mathcal{R}_n \to I$. Since points $y \in \mathcal{R}_n$ have $card\{\mathcal{O}_F(y)\} \le n$ while $card\{\mathcal{O}_G(z)\} > n$ for $z \in Flat(F,\mathcal{P})$, it follows that $h(\mathcal{R}_n) \subset \mathcal{R}_n$. But then h is an injection of the finite set $\mathcal{R}_n$ into itself, hence $h = id$. In particular, $\widetilde{\mathcal{P}} = h(\mathcal{P}) = \mathcal{P}$. $\blacksquare$

Using 1.18 and 1.19 we can show that for certain $\theta \in \mathfrak{P}$ there is no "canonical" map exhibiting the patterns η forced by θ.

1.20. PROPOSITION. *Suppose F is θ-adjusted on $\mathcal{P}$ and J is a flat $\mathcal{P}$-interval for which the point $p \in \mathcal{P}$ defined by $F(J) = \{p\}$ is the limit from both sides of a sequence of distinct periodic points p_i of F. Then for any continuous map f exhibiting θ, there exist patterns η exhibited by f but not forced by θ.*

PROOF: First, it follows from 1.18 that F itself exhibits patterns not forced by θ. To see this, adjoin to $\mathcal{P}_1$ any point interior to J to obtain a pattern η in which p has one more pre-image than it does in $\mathcal{P}_1$. Now consider the map G constructed in 1.18. For $k > 1 + card\{\mathcal{O}_F(p)\}$, G_k has no point in int J mapping to p. Thus every preimage of p under G_k lies in $\mathcal{P}_1$, and hence p has fewer preimages than required for η. But also, by 1.19, in any map G that agrees with F off $Flat(F,\mathcal{P})$, the point in η corresponding to p for F must also be p for G. Thus η is exhibited by F but not by G_k, so $\theta \not\Rightarrow \eta$.

Second, we see that G_k exhibits a pattern not exhibited by F (hence not forced by θ). Since $G_k(x) \ne F(x)$ for $x \in Flat(F,\mathcal{P})$ the G-image of J must contain a one-sided neighborhood of p. But this contains periodic points other than p. Pick one $x \in$ int J with $f(x) = p_i \ne p$ periodic, and let η be obtained by adjoining all points of period $\le per(p_i)$ and $\mathcal{O}(x)$ to $\mathcal{P}_k$ $(k > per(x) + tail(x))$. Then again, the only representative of $\mathcal{P}_k$ in F is $\mathcal{P}_k$, and the $\mathcal{P}_k$-interval containing p_i must contain a point representing p_i in F–which must be p_i. But then J contains the G-preimage x of p_i in the representative of η in G, but no F-preimage. Hence F fails to exhibit η. Note that the only property of G we used in the preceding argument is that it agrees with F on $\mathcal{P}_k$ and has an interior point of J whose image is periodic of period $< k$.

Now consider any continuous map f exhibiting θ on a set, which we can assume without loss of generality to be $\mathcal{P}$. Assume that f exhibits only patterns forced by θ; we will, ultimately, derive a contradiction from this assumption.

Our assumption on f is a restriction on the set of all preperiodic points of f. To make this restriction clearer, we note that for each integer k, application of 1.14 to $Q = \mathcal{R}_k$ ($\mathcal{R}_k$ as in 1.17) gives a representative $\widetilde{\mathcal{R}}_k$ in f of the pattern exhibited by F on $\mathcal{P} \cup \mathcal{R}_k$. This set must contain all preperiodic points of f whose orbit has cardinality $\le k$, since otherwise f exhibits a pattern not exhibited by the map G_k (and hence not forced by θ). It follows that the ordered conjugacy

$$h_k : \mathcal{P} \cup \mathcal{R}_k \to \widetilde{\mathcal{R}}_k$$

is unique, and hence h_k equals the restriction of h_{k+1} to $\mathcal{P} \cup \mathcal{R}_k$. Now, let

$$\mathcal{R} = \mathcal{P} \cup \bigcup_k \mathcal{R}_k, \quad \widetilde{\mathcal{R}} = \bigcup_k \widetilde{\mathcal{R}}_k.$$

The h_k's define an ordered conjugacy

$$h : \mathcal{R} \to \widetilde{\mathcal{R}}.$$

Note that $\mathcal{R}$ contains all preperiodic points of F outside $Flat(F, \mathcal{P})$, and so $\widetilde{\mathcal{R}}$ contains all preperiodic points of f.

We will use this to show that p must be a limit from both sides of a sequence of distinct periodic points for f (as it is for F). To this end, we first prove

CLAIM :*Suppose* $I = \text{clos}\langle x_1, x_2 \rangle$ *is a closed interval satisfying:*

(i) *I contains a unique preperiodic point, x, which is periodic;*

(ii) *each endpoint x_1, x_2 is the limit of a sequence of distinct periodic points of f.*

Then $x_1 = x_2 = x$.

To see the claim, suppose $f^k(x) = x$. Note first that $f^k[I] \subset I$, since otherwise $f^k[I]$ contains infinitely many distinct periodic points, contradicting (i). Similarly, $f^k(x_i) \in \text{int} I$ for $i = 1$ or 2 would also contradict (i), so $f^k\{x_1, x_2\} \subset \{x_1, x_2\}$. This makes both x_1 and x_2 preperiodic, so $x_1 = x_2 = x$, as required by the claim.

Now let $p_i \uparrow p$, $q_i \downarrow p$ be sequences of distinct periodic points for F. Let $\widetilde{p}_i = h(p_i)$, $\widetilde{q}_i = h(q_i)$. Monotonicity of h gives

$$\widetilde{p}_i \uparrow \widetilde{p}_\infty, \quad \widetilde{q}_i \downarrow \widetilde{q}_\infty, \quad \text{and} \quad \widetilde{p}_\infty \leq p \leq \widetilde{q}_\infty.$$

We claim $\widetilde{p}_\infty = \widetilde{q}_\infty$. Let $\widetilde{I} = [\widetilde{p}_\infty, \widetilde{q}_\infty]$ and note that

$$(*) \qquad\qquad\qquad \widetilde{I} \cap \widetilde{\mathcal{R}} = \{p\}.$$

If p is periodic, the earlier claim with $x_1 = \widetilde{p}_\infty$, $x = p$, $x_2 = \widetilde{q}_\infty$ gives the desired equality. If p is not periodic, let $x = f^t(p)$ be a periodic image of p, and let $x_1 = f^t(\widetilde{p}_\infty)$. Then $f^t(\widetilde{q}_\infty)$ cannot be strictly between x_1 and x, because then $\langle x_1, x \rangle$ contains periodic points $f^t(\widetilde{q}_i)$ for i large, contradicting $(*)$. Similarly, x_1 cannot be strictlty between $f^t(\widetilde{q}_\infty)$ and x. Let $x_2 = x$ if $f^t(\widetilde{q}_\infty) = x_1$, and $x_2 = f^t(\widetilde{q}_\infty)$ otherwise. Then $\text{clos}\langle x_1, x_2 \rangle$ satisfies the hypotheses of our earlier claim, and hence $f^t(\widetilde{q}_\infty) = f^t(\widetilde{p}_\infty) = f^t(p)$. Since p is preperiodic, so are $\widetilde{q}_\infty$ and $\widetilde{p}_\infty$, and hence they are all equal, as required.

Finally, to derive our contradiction, we consider the image of J under f. If $f(J) = \{p\}$, f exhibits η, while if $f(J)$ contains more than p, it contains a periodic point, say of period k, not in $\mathcal{P}$. Now, we also know from *1.14* that f exhibits θ_k on a set $\mathcal{P}'_k$ containing $\mathcal{P}$. Hence by an argument like that for G, f exhibits another pattern not forced by θ. ∎

1.20 shows that we cannot in general expect to model the set of patterns η with $\theta \Rightarrow \eta$ by a criterion as simple as "$\theta \Rightarrow \eta$ iff F_θ exhibits η", where F_θ is some canonically constructed map (possibly different from the θ-adjusted maps). However, the following result shows that we *can* use θ-adjusted maps to decide which patterns η exhibited by F satisfy $\theta \Rightarrow \eta$.

1.21. THEOREM. *Let F be θ-adjusted on $\mathcal{P}$, $\theta \in \mathfrak{P}$, and let $\eta \in \mathfrak{P}$. Then $\theta \Rightarrow \eta$ if and only if there is a representative $\mathcal{Q}$ of η in F with $\mathcal{Q} \cap Flat(F, \mathcal{P}) = \emptyset$.*

PROOF: It follows from *1.14* that if $\mathcal{Q} \cap Flat(F, \mathcal{P}) = \emptyset$ and $\mathcal{Q}$ represents η in F, then $\theta \Rightarrow \eta$. Conversely, suppose that $\theta \Rightarrow \eta$. Let G_k be the map constructed from F in *1.18*, with k the degree of η. Then η has a representative $\mathcal{Q}$ in G. Since every point in $\mathcal{Q}$ has an orbit of cardinality k or less, it follows from *1.18* that $\mathcal{Q} \cap Flat(F, \mathcal{P}) = \emptyset$. But $F = G$ off $Flat(F, \mathcal{P})$, so $\mathcal{Q}$ is a representative of η in F. ∎

2. Equivalence and Essential Patterns

In this section, we use some of the results in §1 to study the extent to which the forcing relation partially orders the set of patterns.

Baldwin [**Ba**] showed that forcing partially orders the set of cycles. Since forcing is trivially transitive and reflexive, the only issue is antisymmetry: for $\theta, \eta \in \mathfrak{C}$, Baldwin showed that $\theta \Rightarrow \eta \Rightarrow \theta$ implies $\theta = \eta$. This is easily seen to be false if $\eta \notin \mathfrak{C}$ and θ strongly forces some cycle: for such θ, take F θ-adjusted on $\mathcal{P}$ and let $\mathcal{Q}$ be any periodic orbit of F disjoint from $\mathcal{P}$. Then the pattern η represented by $\mathcal{P} \cup \mathcal{Q}$ contains θ as a subpattern, so $\eta \Rightarrow \theta$; but since $[\mathcal{P} \cup \mathcal{Q}] \cap Flat(F, \mathcal{P}) = \emptyset$, we have $\theta \Rightarrow \eta$ as well, by *1.14*. Note that if $\theta \in \mathfrak{S}$, this gives $\eta \in \mathfrak{S}$, so antisymmetry fails even for permutations.

Let us call $\theta, \eta \in \mathfrak{P}$ **equivalent** if $\theta \Rightarrow \eta \Rightarrow \theta$. Note that this is a true equivalence relation. We shall see that the various elements η of a single equivalence class arise from an *essential* subpattern θ by a process like that illustrated above. Given any pattern, the (unique) essential pattern equivalent to it will be easily computed.

Suppose $\theta \in \mathfrak{P}_n$, and let $\mathcal{P} = \{1, \ldots, n\}$. We adopt the following terminology, related to that in [**BlCv1**]. The points $1, n$ are the **endpoints** of θ. A non-endpoint $i \in \{2, \ldots, n-1\}$ is a **turning point** if $\theta(i) >_\sigma \theta(i-1), \theta(i+1)$, where $\sigma = \pm 1$; we say i is a **maximum** or **minimum** as $\sigma = +1$ or $\sigma = -1$. A **flat block** for θ is a set $\{i, \ldots, i+s\}$ of consecutive elements of $\mathcal{P}$ for which $\theta(i) = \theta(i+1) = \cdots = \theta(i+s)$. A **tandem block** for θ is a set $\{i, \ldots, i+s\}$ of consecutive elements of $\mathcal{P}$ with $i+j$ and $i+j+1$ in tandem cycles of θ for $j = 0, \ldots, s-1$. Note that if F is θ-adjusted on $\mathcal{P}$ then the convex hull of any maximal flat block is a maximal flat interval for F, and the turning points of θ of maximum (resp. minimum) type are precisely the strict relative maxima (resp. minima) of F. Of course, F may also have degenerate extrema on flat intervals; accordingly we define a **turning element of** θ of maximum (resp. minimum) type to be a block $\{i, \ldots, i+s\}$ of $s+1$ (possibly $s = 0$) consecutive non-endpoint elements of $\mathcal{P}$ such that

$$\theta(i-1) <_\sigma \theta(i) = \theta(i+1) = \cdots = \theta(i+s) >_\sigma \theta(i+s+1).$$

A turning element with $s = 0$ is a turning point. If F is θ-monotone, the convex hull of a turning element of θ will be called a **turning element for** F. Finally, a **critical point** for θ is any endpoint, turning point, or member of a flat block for θ. For any θ-adjusted map F, the critical points of θ are precisely the points outside $Flat(F, \mathcal{P})$ which are critical for F in the sense that they are not interior to any interval on which F is strictly monotone.

In the following, we find it convenient to use the notation $A < B$ for two sets to mean that $a < b$ for any $a \in A$ and $b \in B$.

2.1. LEMMA. *Given $\theta, \eta \in \mathfrak{P}$ with $\theta \Rightarrow \eta$, suppose F is θ-monotone on $\mathcal{P}$ and $\mathcal{Q}$ is a representative of η in F with $\mathcal{Q} \cap Flat(F, \mathcal{P}) = \emptyset$.*

(i) *Suppose*

$$y_1 < y_2 < \cdots < y_k \in \mathcal{Q}$$

is a choice of one point from each turning element for η, and the turning element containing y_i is of σ_i-maximal type. Then it is possible to pick

$$x_1 < x_2 < \cdots < x_k \in \mathcal{P}$$

so that x_i belongs to a turning element for θ of σ_i-maximal type, such that F is monotone on $\langle x_i, y_i \rangle$,

$$F(x_i) \geq_{\sigma_i} F(y_i),$$

and

$$\{x_i, y_i\} < \{x_{i+1}, y_{i+1}\}$$

for $i = 1, \ldots, k-1$.

(ii) *If $p, q \in \mathcal{Q}$ with $(p, q) \cap \mathcal{Q} = \emptyset$ and $F(p) = F(q)$, then either $p, q \in \mathcal{P}$ belong to a single flat block for θ or the open interval (p, q) contains a turning element for F.*

PROOF:

Proof of (i): Each y_i belongs to a maximal closed interval $[a_i, b_i]$ on which F is (weakly) monotone. Precisely one of the endpoints of this interval belongs to a turning element for F of σ_i-maximal type; let this be x_i. Clearly,

(*) $F(x_i) \geq_{\sigma_i} F(y_i).$

Since the y_i's include a point in every turning element for η, we have $\sigma_{i+1} = -\sigma_i$ for $i = 0, \ldots, k-1$. Thus $b_i \leq a_{i+1}$ unless $a_i = a_{i+1}$ and $b_i = b_{i+1}$, and it is easy to see using the inequality (*) that in either case we have

$$\{x_i, y_i\} < \{x_{i+1}, y_{i+1}\}.$$

Proof of (ii): If $[p, q]$ is a flat interval for F, then since $p, q \notin Flat(F, \mathcal{P})$ we must have $p, q \in \mathcal{P}$, and $[p, q] \cap \mathcal{P}$ is a flat block for θ containing p and q. If $[p, q]$ is not flat, then some interior point of $[p, q]$ achieves a σ-maximum ($\sigma \in \{\pm 1\}$) for F on $[p, q]$; this value is then achieved on a turning element for F, interior to $[p, q]$. ∎

Using *1.18, 1.19, 1.21* and *2.1*, we obtain the following basic observation.

2.2. PROPOSITION. *Equivalent patterns have the same critical points. More precisely, suppose $\theta \Rightarrow \eta \Rightarrow \theta$, let F be θ-adjusted on $\mathcal{P} = \{1, \ldots, n\}$ and suppose $\mathcal{Q} = \{y_1, \ldots, y_n\}$ is a representative of η in F with $\mathcal{Q} \cap Flat(F, \mathcal{P}) = \emptyset$. Let G be η-adjusted on $\mathcal{Q}$. Then:*

(i) *$\mathcal{P}$ and $\mathcal{Q}$ have the same endpoints: $y_1 = 1$ and $y_k = n$;*

(ii) *$Flat(F, \mathcal{P}) = Flat(G, \mathcal{Q})$; in particular, the flat blocks of θ (as points in $\mathcal{P} \subset [1, n]$) coincide with the flat blocks of η (as points in $\mathcal{Q} \subset [1, n]$);*

(iii) *G is $\mathcal{P}$-monotone: the turning points of θ (as points in $\mathcal{P} \subset [1, n]$) and of η (as points in $\mathcal{Q} \subset [1, n]$) coincide.*

PROOF: We are given F θ-adjusted on $\mathcal{P}$ and G η-adjusted on $\mathcal{Q}$, with $\mathcal{Q} \cap Flat(F, \mathcal{P}) = \emptyset$. Since $\eta \Rightarrow \theta$, by *1.21* there is a representative $\mathcal{P}'$ of θ in G, with $\mathcal{P}' \cap Flat(G, \mathcal{Q}) = \emptyset$; note that $\mathcal{P}'$ is contained in the convex hull of $\mathcal{Q}$. Since G is η-adjusted on $\mathcal{Q}$ and F exhibits η on $\mathcal{Q}$, the identity map on $\mathcal{Q}$ extends to an ordered conjugacy h from $\mathcal{Q} \cup \mathcal{P}'$ to $\mathcal{Q} \cup \mathcal{P}''$, where $\mathcal{P}''$ is a representative of θ in F, contained in the convex hull of $\mathcal{Q}$. Using *1.18* as in the proof of *1.20*, we can assume $\mathcal{P}'' \cup Flat(F, \mathcal{P}) = \emptyset$. But by *1.19*, this means $\mathcal{P}'' = \mathcal{P}$. In particular, $\mathcal{P}$ is contained in the convex hull of $\mathcal{Q}$; but the reverse inclusion is trivial, so the two convex hulls are the same, proving *(i)*.

Furthermore, since $\mathcal{P}' \cap Flat(G, \mathcal{Q}) = \emptyset$, it follows from the existence of the ordered conjugacy $h : \mathcal{Q} \cup \mathcal{P}' \to \mathcal{Q} \cup \mathcal{P}$ that $\mathcal{P} \cap Flat(G, \mathcal{Q}) = \emptyset$ (since $Flat(G, \mathcal{Q})$ is determined by the flat blocks in $\mathcal{Q}$). Suppose now that y_i, y_{i+1} belong to a flat block in $\mathcal{Q}$; this means $F(y_i) = F(y_{i+1})$ and lemma *2.1(ii)* applies; but case *(b)* is impossible because $\mathcal{P} \cap Flat(G, \mathcal{Q}) = \emptyset$. Thus, y_i, $y_{i+1} \in \mathcal{P}$ belong to a single flat block in $\mathcal{P}$. This argument applies to any pair of adjacent elements of a flat block in $\mathcal{Q}$, so that every flat block in $\mathcal{Q}$ is contained in a flat block in $\mathcal{P}$, and $Flat(G, \mathcal{Q}) \subset Flat(F, \mathcal{P})$. A symmetric argument gives the opposite inclusion, proving *(ii)*.

Finally, let $y_i \in \mathcal{Q}$ be a turning point for G. Then (y_{i-1}, y_{i+1}) contains no flat intervals for F (by *(ii)* above) but does contain a turning element for F (by *2.1(i)*), which must therefore be a turning point, z, of F, of the same type as y_i. We wish to show $z = y_i$.

Suppose $z \neq y_i$. We can assume z is picked so that F has no turning points between z and y_i; let J be the $\mathcal{Q}$-interval $((y_{i-1}, y_i)$ or $(y_i, y_{i+1}))$ containing z. Then y_i is a σ-maximum for G on (y_{i-1}, y_i), while z is a σ-maximum for F on $\mathrm{clos}\langle z, y_i \rangle$. Note that $G(x) <_\sigma G(y)$ for $x \in \mathrm{int}\, J$. Now, z and the endpoints of J belong to $\mathcal{P} \cup \mathcal{Q}$, and $F(z)$ is not between the F-images of these two endpoints. Using the ordered conjugacy, this gives us $z' \in \mathcal{P}' \cap \mathrm{int}\, J$ with $G(z') >_\sigma G(y_i)$, contradicting our earlier inequality. But this means $z = y_i$. Thus, every critical point of G in $\mathcal{Q}$ is also a critical point for F, hence in $\mathcal{P}$. The reverse inequality holds by a symmetric argument, and proves *(iii)*. ∎

We turn now to considering tandem blocks for equivalent patterns. Recall that in *1.4* we constructed a pattern $\theta_{\frac{1}{2}}$ equivalent to any given $\theta \in \mathfrak{P}$ such that $\theta_{\frac{1}{2}}$ has no self-tandem cycles. Our descriptions are simplified if we replace each pattern θ by the equivalent pattern $\theta_{\frac{1}{2}}$ without self-tandem cycles, or (equivalently) restrict our attention to patterns without self-tandem cycles.

2.3. LEMMA. *Suppose θ and η are equivalent patterns and neither has self-tandem cycles; let F be θ-adjusted on $\mathcal{P} = \{1, \dots, n\}$ and let $\mathcal{Q} = \{y_1, \dots, y_k\}$ be a representative of η in F with $\mathcal{Q} \cap Flat(F, \mathcal{P}) = \emptyset$. If $i, i+1$ belong to tandem cycles in η, then $y_i, y_{i+1} \in \mathcal{P}$ and belong to a tandem block in θ.*

PROOF: Let $J = [y_i, y_{i+1}]$ and set $J_j = F^j[J]$; assume $J_0, \dots, J_{\ell-1}$ have disjoint interiors and $J_\ell = J_0$. Since J_j is a $\mathcal{Q}$-interval, it contains no critical points of η in its interior; by *2.3(iii)*, this means F has no critical points interior to J_j, and F is a homeomorphism on J_j for all j. In particular, since the endpoints of J_0 are fixed under F^ℓ, any pre-periodic point interior to J_0 is actually periodic with period ℓ. Thus, $\mathcal{P} \cap J_0$ consists only of periodic points in a single tandem block of θ. If $y_i, y_{i+1} \in \mathcal{P}$, we are done. If not, apply *1.8* to $x = y_i$ or y_{i+1}, whichever does not belong to $\mathcal{P}$, with $k = \ell$ to conclude that x is interior to a self-tandem interval. Since by hypothesis θ has no self-tandem blocks, $y_i, y_{i+1} \in \mathcal{P}$ and we are done. ∎

As a corollary, we obtain the following analogue for tandem blocks of *2.2*.

2.4. PROPOSITION. *Suppose θ, η are equivalent patterns, both without self-tandem cycles; let F be θ-adjusted on $\mathcal{P} = \{1, \ldots, n\}$ and pick $\mathcal{Q} = \{y_i, \ldots, y_k\}$ an η-representative in F with $\mathcal{Q} \cap Flat(F, \mathcal{P}) = \emptyset$. Then the tandem blocks of θ (as subsets of $\mathcal{P}$) coincide with the tandem blocks of η (as subsets of $\mathcal{Q}$).*

PROOF: By *2.3*, every maximal tandem block for η in $\mathcal{Q}$ is actually a subset of $\mathcal{P} \cap \mathcal{Q}$ and is contained in a single maximal tandem block for θ. Conversely, if G is η-adjusted on $\mathcal{Q}$, and $\mathcal{P}'$ is a θ-representative in G, then every maximal tandem block for θ (in $\mathcal{P}'$) actually belongs to $\mathcal{Q}$, and is contained in a single tandem block for η; by *1.4*, since F exhibits η on $\mathcal{Q}$, this last statement holds for F, as well. But we now have inclusion both ways, showing equality of the two sets of tandem blocks. ∎

2.2 and *2.4* give us most of the tools necessary for deciding whether two patterns are equivalent. The following limited version of a result of Block and Coven **[BlCv1]** is our final tool.

2.5. LEMMA. [BlCv1] *Given $\theta \in \mathfrak{P}_n$, let $\mathcal{P}^* \subset \{1, \ldots, n\}$ consist of the endpoints $1, n$, the turning points of θ, and the endpoints of all maximal flat or tandem blocks in θ, together with the orbits of all these points, and denote by θ^* the subpattern of θ represented by $\theta|\mathcal{P}^*$.*

The the following are equivalent for $\theta, \eta \in \mathfrak{P}$:

(i) $\theta^* = \eta^*$;

(ii) there exists an ordered conjugacy from any θ-linear map to any η-linear map.

Note that patterns with the same linear map need not be equivalent in our sense, because the flat or tandem blocks may differ in size. We will see that this is essentially the only difference between ordered conjugacy of linear maps and equivalence of patterns.

2.6. DEFINITION. *Given $\theta \in \mathfrak{P}_n$, we define two patterns related to θ:*

*(i) Form $\theta_{\frac{1}{2}}$ as in 1.4, and let $\mathcal{P}_* \subset \mathcal{P}_{\frac{1}{2}}$ consist of all critical points of $\theta_{\frac{1}{2}}$, all elements of tandem blocks in $\theta_{\frac{1}{2}}$, and the forward orbits of all these points. The **fundamental pattern** θ_* is represented by the restriction of $\theta_{\frac{1}{2}}$ to $\mathcal{P}_*$.*

*(ii) For any (maximal) tandem block of θ_*, there is at most one element of the block whose period differs from (in fact, equals half of) the period of each of the ends of the block; call this the **center** of the block, if it exists. The **essential subpattern** θ_{**} is formed by deleting the centers from all tandem blocks of θ_* that possess one.*

2.7. REMARK. *It is easy to see that for any $\theta \in \mathfrak{P}$*

(i) $\theta_* = (\theta_{**})_{\frac{1}{2}} = (\theta_{\frac{1}{2}})_*$

(ii) $(\theta_{**})_{**} = (\theta_*)_{**} = (\theta_{\frac{1}{2}})_{**} = \theta_{**}$;

in particular, θ_* is a subpattern of $\theta_{\frac{1}{2}}$ and θ_{**} is a subpattern of θ.

The main result of this section is:

2.8. THEOREM. *Given $\theta, \eta \in \mathfrak{P}$, the following are equivalent:*

(i) $\theta \Rightarrow \eta \Rightarrow \theta$;
(ii) $\theta_* = \eta_*$;
(iii) $\theta_{**} = \eta_{**}$.

PROOF: By Remark *2.7.* since θ is equivalent to $\theta_{\frac{1}{2}}$, we can assume $\theta = \theta_{\frac{1}{2}}$ and $\eta = \eta_{\frac{1}{2}}$ in our proof.

(i) implies (ii): Given $\theta \Rightarrow \eta \Rightarrow \theta$, let F be θ-adjusted on $\mathcal{P}$ and let $\mathcal{Q}$ be a representative of η in F with $\mathcal{Q} \cap Flat(F, \mathcal{P}) = \emptyset$. Under the assumption above, *2.2* and *2.4* tell us that the sets $\mathcal{P}_*$ and $\mathcal{Q}_*$ (with the obvious meaning) coincide, so $\theta_* = \eta_*$.

(ii) implies (iii): Obvious.

(iii) implies (i): Note that the pattern θ^* in *2.5* is actually a subpattern of θ_{**}, and $\theta_{**} = \eta_{**}$ implies $\theta_* = \eta_*$. Thus by *2.5*, *(iii)* implies that any θ-linear map is topologically conjugate (via an ordered conjugacy) to any η-linear map. Let f be θ-linear on $\mathcal{P} = \{1, \ldots, n\}$; if g is η-linear and h is an ordered conjugacy from f to g, let $\mathcal{Q}$ be the h-image of the standard η-representative in g. $\mathcal{Q}$ represents η in f. Since h is a conjugacy, critical points of g map to critical points of f; thus the critical points in $\mathcal{Q}$ and $\mathcal{P}$ coincide; similarly, h takes flat (resp. tandem) intervals for g to flat (resp. tandem) intervals for f. Since $\theta_* = \eta_*$, the number of elements of $\mathcal{Q}$ in a maximal flat (resp. tandem) interval for f is the same as the number of points of $\mathcal{P}$ in that interval. Thus, we can adjust $\mathcal{Q}$ so that these points coincide. But then $\mathcal{Q} \cap Flat(f, \mathcal{P}) = \emptyset$. Form the canonical θ-adjusted map F (see *1.7*). Then F is θ-adjusted, and $F|\mathcal{Q} = f|\mathcal{Q}$, so $\mathcal{Q}$ is a representative of η in F with $\mathcal{Q} \cap Flat(F, \mathcal{P}) = \emptyset$. It follows from *1.21* that $\theta \Rightarrow \eta$. A symmetric argument gives $\eta \Rightarrow \theta$, and completes the proof. ∎

We close this section with a few remarks concerning the essential subpattern θ_{**} of $\theta \in \mathfrak{P}$. At the beginning of this section, we saw that given any pattern θ strongly forcing something, we could form a new pattern $\eta \neq \theta$ equivalent to θ by adjoining some orbits of a θ-adjusted map to $\mathcal{P}$; in this case θ is a proper subpattern of η such that $\theta \Rightarrow \eta$. Let us call a pattern **redundant** if it is forced by (hence equivalent to) some proper subpattern, and **essential** if it is not redundant. An easy corollary of *2.8* is

2.9. COROLLARY.

(i) $\theta \in \mathfrak{P}$ is essential if and only if $\theta = \theta_{**}$;
(ii) the forcing relation is a partial ordering on the set of all essential patterns (resp. permutations).

PROOF:

Proof of (i): If $\eta \subseteq \theta$ is a proper subpattern forcing θ, then $\eta_{**} = \theta_{**}$; since $\eta_{**} \subseteq \eta$, this says $\theta \neq \theta_{**}$; conversely, since θ_{**} is a subpattern forcing θ, if θ is essential then θ_{**} cannot be a proper subpattern, so that $\theta = \theta_{**}$.

Proof of (ii): As remarked at the beginning of this section, the only issue is antisymmetry, and this is an obvious corollary of *2.8* and *2.9(i)*. ∎

3. REDUCTIONS AND EXTENSIONS OF PATTERNS

The notion of an extension of a cyclic permutation was formulated in [**ALM**], although special cases of this idea (under various pseudonyms) have been useful in other contexts [**ALS;BlC;Be1;C;H1-2**]. In this section, we formulate a general definition of extension in the context of patterns and relate it to forcing. We shall see in the next section that it is also related to properties of the *Markov graph*.

Let $\mathcal{P} = \{1, \ldots, n\}$. Recall that by a **block** in $\mathcal{P}$, we mean a set of the form $B = \{i \in \mathcal{P} \mid a \leq i \leq b\}$, where $a \leq b \in \mathcal{P}$.

3.1. DEFINITION. *Suppose $\theta \in \mathfrak{P}$. By a **block structure** for θ, we mean a partition $\mathcal{B} = \{B_1, \ldots, B_k\}$ of $\mathcal{P}$ into disjoint blocks such that if $x, y \in \mathcal{P}$ belong to the same block, their images under θ belong to a single block. We number the blocks B_j so that $x \in B_i$, $y \in B_j$ and $i < j$ implies that $x < y$; then there is a unique pattern $\eta \in \mathfrak{P}_k$ defined by*

$$\theta[B_j] \subset B_{\eta(j)}, \qquad j = 1, \ldots, k.$$

*We call η a **reduction** of θ, and θ an **extension** of η, and refer to $\mathcal{B}$ as a **block structure** for θ over η.*

Several observations are in order; the first three below are immediate.

3.2. REMARK.

(i) *Every pattern $\theta \in \mathfrak{P}_n$ has two **trivial reductions**: $\eta = \theta$ $(k = n)$ and η a single fixedpoint $(k = 1)$;*
(ii) *if $\theta \in \mathfrak{S}$, then any reduction η of θ belongs to $\mathfrak{S}$;*
(iii) *if $\theta \in \mathfrak{C}_n$, then any reduction η of θ belongs to $\mathfrak{C}_k$, for some divisor k of m; if $n = mk$ then each block of $\mathcal{B}$ has m elements.*
(iv) *While the reduction η is uniquely determined by the block structure $\mathcal{B}$, the converse is not true. However, if θ(hence η) is a cycle, then there is at most one block structure for θ over η.*

PROOF:

Proof of (iv): For the first comment, we provide two examples. The first (figure 3) is $\theta \in \mathfrak{P}_3$, where

$$\theta(1) = \theta(2) = \theta(3) = 3.$$

Then

$$B_1 = \{1, 2\}, \ B_2 = \{3\}$$

is a block structure over $\eta \in \mathfrak{P}_2$, where $\eta(1) = \eta(2) = 2$. However, another block structure over η is given by

$$\widetilde{B}_1 = \{1\}, \ \widetilde{B}_2 = \{2, 3\}.$$

27

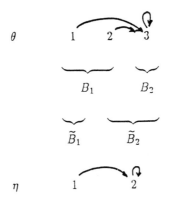

Figure 3

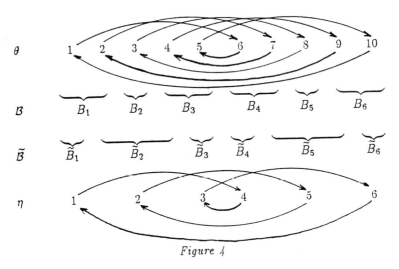

Figure 4

A second example (figure 4) is $\theta \in \mathfrak{S}_{10}$, where

$$\theta = (1\ 6\ 5\ 10)(2\ 7\ 4\ 9)(3\ 8).$$

One block structure is

$$B_1 = \{1,2\},\ B_2 = \{3\},\ B_3 = \{4,5\},\ B_4 = \{6,7\},\ B_5 = \{8\},\ B_6 = \{9,10\}$$

with reduction

$$\eta = (1\ 4\ 3\ 6)(2\ 5),$$

while another block structure over η is

$$\widetilde{B}_1 = \{1\},\ \widetilde{B}_2 = \{2,3,4\},\ \widetilde{B}_3 = \{5\},\ \widetilde{B}_4 = \{6\},\ \widetilde{B}_5 = \{7,8,9\},\ \widetilde{B}_6 = \{10\}.$$

The uniqueness of the block structure for a cycle θ over a cycle η follows immediately from *(iii)*. ∎

In the case of reductions of cycles, we introduce a more precise terminology: if $\theta \in \mathfrak{C}_{mk}$ is an extension of $\eta \in \mathfrak{C}_k$, we call θ an *m*-**extension** of η and η an

m-**reduction** of θ. Of particular importance in our theory is the case $m = 2$; we refer to a 2-extension of a cycle as a **doubling** of it.

Given a block structure $\mathcal{B} = \{B_j\}_{j=1}^k$ for $\theta \in \mathfrak{P}_n$ and a representative $\mathcal{P} = \{x_1 < x_2 < \cdots < x_n\}$ for θ, we define $\mathcal{B}$-**intervals** with respect to $\mathcal{P}$ as follows. Given a block $B_j \in \mathcal{B}$, we have $B_j = \{i \in \mathcal{P} \mid a_j \le i \le b_j\}$ for some $a_j \le b_j$ in $\mathcal{P}$; define the jth $\mathcal{B}$-interval as

$$\mathcal{R}(B_j, \mathcal{P}) = [x_{a_j}, x_{b_j}].$$

Note that $\mathcal{R}(B_j, \mathcal{P})$ consists either of a single point of $\mathcal{P}$ (if $a_j = b_j$) or the union of all $\mathcal{P}$-intervals whose endpoints both correspond to elements of the block B_j; these $\mathcal{B}$-intervals are disjoint closed intervals. The **representative set** for $\mathcal{B}$ with respect to $\mathcal{P}$ is

$$\mathcal{R}(\mathcal{B}, \mathcal{P}) = \bigcup_{B_j \in \mathcal{B}} \mathcal{R}(B_j, \mathcal{P}).$$

The $\mathcal{B}$-intervals are the components of the representative set $\mathcal{R}(\mathcal{B}, \mathcal{P})$.

3.3. LEMMA. *Suppose* $\mathcal{B} = \{B_j\}$ *is a block structure for* $\theta \in \mathfrak{P}$ *over* $\eta \in \mathfrak{P}$, *and suppose* f *is* θ-*monotone with respect to* $\mathcal{P}$. *Then:*

  (i) $f[\mathcal{R}(B_j, \mathcal{P})] \subset \mathcal{R}(B_{\eta(j)}, \mathcal{P})$;
  (ii) *if* $\mathcal{Q}$ *is a finite* f-*invariant set representing* $\xi \in \mathfrak{P}$ *and* $\mathcal{Q} \subset \mathcal{R}(\mathcal{B}, \mathcal{P})$, *then the sets* $\mathcal{R}(B_j, \mathcal{P})$ *induce a block structure for* ξ *over a subpattern of* η.

PROOF: Let k denote the number of blocks in $\mathcal{B}$, so $deg(\eta) = k$.

Proof of (i): Fix j. If i, $i+1 \in B_j$, then by definition of block structure, $\theta(i)$, $\theta(i+1) \in B_{\eta(j)}$; thus by θ-monotonicity of f, $f[x_i, x_{i+1}] \subset \mathcal{R}(B_{\eta(j)}, \mathcal{P})$. Taking the union over i such that i, $i+1 \in B_j$, we get *(i)*.

Proof of (ii): Suppose $\mathcal{Q} = \{y_1 < y_2 < \cdots < y_m\}$ so $f(y_j) = y_{\xi(j)}$. Define

$$B_j(\mathcal{Q}) = \{i \in \{1, \ldots, m\} \mid y_i \in \mathcal{R}(B_j, \mathcal{P})\}.$$

Then each $B_j(\mathcal{Q})$ is a block and by *(i)*, $i \in B_j(\mathcal{Q})$ implies $\xi(i) \in B_{\eta(j)}(\mathcal{Q})$. In particular, the restriction of η to $\{j \in \{1, \ldots, k\} \mid B_j(\mathcal{Q}) \ne \emptyset\}$ represents a subpattern of η and a reduction of ξ. ∎

Now consider invariant sets disjoint from the representative set $\mathcal{R}(\mathcal{B}, \mathcal{P})$. If i, $i+1$ belong to distinct $\mathcal{B}$-intervals, then the open $\mathcal{P}$-interval (x_i, x_{i+1}) is a $\mathcal{B}$-**gap**; note that if $x_i \in B_j$ then $x_{i+1} \in B_j \cup B_{j+1}$, so that a $\mathcal{B}$-gap is a component of the complement of $\mathcal{R}(\mathcal{B}, \mathcal{P})$. Suppose now that F is θ-adjusted on $\mathcal{P} = \{x_1, \ldots, x_n\}$ and g represents $\eta \in \mathfrak{P}_k$ on $\mathcal{P}_\mathcal{B} = \{y_1, \ldots, y_k\}$, where $\mathcal{B} = \{B_j\}_{j=1}^k$ is a block structure for θ with reduction η. Define $\pi : \mathcal{P} \to \mathcal{P}_\mathcal{B}$ by

$$\pi(x_i) = y_j, \quad \text{where } i \in B_j.$$

Note that if int $J_i = (x_i, x_{i+1})$ is a $\mathcal{B}$-gap then $\pi[J_i] = [\pi(x_i), \pi(x_{i+1})]$ is a $\mathcal{P}_\mathcal{B}$-interval, and π is a one-to-one order-preserving correspondence between the $\mathcal{B}$-gaps and the $\mathcal{P}_\mathcal{B}$-intervals.

Now, given $\mathcal{Q} = \{z_1, \ldots, z_\ell\}$ a finite invariant set for F such that

$$\mathcal{Q} \cap [\mathcal{R}(\mathcal{B}, \mathcal{P}) \cup Flat(F, \mathcal{P})] = \emptyset,$$

we apply *1.13* to find N such that the proper F-itineraries of length N with respect to $\mathcal{P}$ separate points of $\mathcal{Q}$. That is, given $z \in \mathcal{Q}$, since $F^j(z) \notin \mathcal{P}$ for all j, there is a unique proper itinerary $\mathcal{I}(z) = \{I_0, \ldots, I_N\}$ such that $z \in \mathcal{R}(F, \mathcal{I}(z))$.

Furthermore, since $\mathcal{Q} \cap \mathcal{R}(\mathcal{B}, \mathcal{P}) = \emptyset$, each I_j is a $\mathcal{B}$-gap. Now, note that if I_i F-covers I_j and both are $\mathcal{B}$-gaps, then $\pi[I_i]$ g-covers $\pi[I_j]$. Thus, the set $\mathcal{R}(g, \pi[\mathcal{I}(z)])$ is nonempty, since $\pi[\mathcal{I}(z)] = \{\pi[I_j]\}_{j=0}^N$ is a proper g-itinerary. It is easy to see that $\pi[\mathcal{I}(z)]$ is a p-loop iff $\mathcal{I}(z)$ is a p-loop; also, if $z_1 < z_2$, then $\mathcal{R}(F, \mathcal{I}(z_1)) < \mathcal{R}(F, \mathcal{I}(z_2))$, and then $\mathcal{R}(g, \pi[\mathcal{I}(z_1)]) < \mathcal{R}(g, \pi[\mathcal{I}(z_2)])$. By a process as in *1.14*, we can pick a point $\pi(z) \in \mathcal{R}(g, \pi[\mathcal{I}(z)])$ for every $z \in \mathcal{Q}$ so that $g(\pi(z)) = \pi(F(z))$. We have proved

3.4. LEMMA. *If $\mathcal{B} = \{B_j\}$ is a block structure for $\theta \in \mathfrak{P}$ with reduction $\eta \in \mathfrak{P}$, and F is θ-adjusted on $\mathcal{P}$, then for any pattern ξ with a representative $\mathcal{Q}$ in F disjoint from $\mathcal{R}(\mathcal{B}, \mathcal{P}) \cup Flat(F, \mathcal{P})$, $\eta \Rightarrow \xi$.*

We note that what one might expect to be a more natural proof of *3.4*, namely to extend π to a semiconjugacy of F with some η-adjusted map g, is impossible.

3.5. EXAMPLE. *let $\theta \in \mathfrak{P}_4$ (figure 5) be defined by*

$$\theta(1) = 1 = \theta(2)$$
$$\theta(3) = 3$$
$$\theta(4) = 2$$

and define the block structure $\mathcal{B} = \{B_j\}_{j=1}^3$ by

$$B_1 = \{1, 2\}, \quad B_2 = \{3\}, \quad B_3 = \{4\}.$$

Let F be θ-linear with respect to $\mathcal{P} = \{1, 2, 3, 4\}$ and g be η-adjusted with respect to $\mathcal{P}_{\mathcal{B}} = \{1, 2, 3\}$, where η is the reduction of θ associated to $\mathcal{B}$. Then

(i) *F is θ-adjusted;*
(ii) *there is no semiconjugacy extending $\pi : \mathcal{P} \to \mathcal{P}_{\mathcal{B}}$;*
(iii) *F does not exhibit η (so $\theta \not\Rightarrow \eta$).*

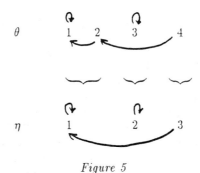

Figure 5

PROOF: Note that $\eta \in \mathfrak{P}_3$ is given by

$$\eta(1) = 1 = \eta(3), \quad \eta(2) = 2,$$

and $\pi : \mathcal{P} \to \mathcal{P}_{\mathcal{B}}$ is

$$\pi(1) = \pi(2) = 1, \quad \pi(3) = 2, \quad \pi(4) = 3.$$

To see *(i)*, note that the only periodic points of F are the two fixedpoints 1 and 3, since *(a)* every point of $[3, 4]$ maps to $[2, 3]$, *(b)* every interior point of $[2, 3]$ has some finite iterate in $[1, 2]$, and *(c)* $F[1, 2] = 1$. Since the periodic points of F are separated by $2 \in \mathcal{P}$ and F is $\mathcal{P}$-monotone, it is $\mathcal{P}$-adjusted. Now, a semiconjugacy extending π must map $[2,3]$ onto $[1,2]$, but g maps the open interval $(1,2)$ to itself while under F every point of $(2,3)$ eventually falls into $[1,2]$. Thus, we cannot extend π and still fulfill $g \circ \pi = \pi \circ F$ on $[2,3]$, showing *(ii)*.

Finally *(iii)* follows from the observation that while both θ and η have precisely two fixedpoints, η has a point (3) to the right of both fixedpoints (1, 2) mapping to the left fixedpoint (1), while in F, no point to the right of the right fixedpoint hits the left fixedpoint. ∎

On the other hand, our arguments give a more positive picture for permutations.

3.6. LEMMA. *Given $\theta \in \mathfrak{P}$, and $\mathcal{B} = \{B_j\}$ a block structure for θ over $\eta \in \mathfrak{P}$, F a θ-adjusted map, and g η-adjusted,*

  (i) *every periodic orbit of F either belongs to or is disjoint from $\mathcal{R}(\mathcal{B}, \mathcal{P})$;*
  (ii) *for every periodic orbit $\mathcal{Q}'$ of g outside $\mathcal{P}_\mathcal{B}$, there exists a unique periodic orbit $\mathcal{Q}$ of F outside $\mathcal{R}(\mathcal{B}, \mathcal{P})$ with $\pi[\mathcal{Q}] = \mathcal{Q}'$.*

PROOF: *(i)* is obvious, since $\mathcal{R}(\mathcal{B}, \mathcal{P})$ is F-invariant.

Proof of (ii): Given a periodic point x for g with $per(x) = \ell$, let $\mathcal{I} = \{I_0, \ldots, I_\ell = I_0\}$ be the unique proper ℓ-loop of $\mathcal{P}_\mathcal{B}$-intervals with x the only fixedpoint of g^ℓ in $\mathcal{R}(g, \mathcal{I})$. Let $J_0, \ldots, J_\ell = J_0$ be the closures of the gaps uniquely associated to $I_0, \ldots, I_\ell$ by π. Then $\mathcal{I} = \{J_0, \ldots, J_\ell\}$ is a proper, non-repetitive ℓ-loop of $\mathcal{P}$-intervals, so there is a unique $y \in J_0$ which is periodic for F with $per(y) = \ell$ and $F^j(y) \in J_j$ for $j = 0, \ldots, \ell - 1$. ∎

Given any pattern $\theta \in \mathfrak{P}_n$, we can define the **permutation part** of θ as the maximum subpermutation **Perm**(θ) of θ: note that $Perm(\theta)$ is the union of all subcycles of θ, and is represented by the restriction of θ to $\theta^n[\mathcal{P}]$, where n is the degree of θ.

3.7. THEOREM. *Suppose $\theta \in \mathfrak{P}$ extends $\eta \in \mathfrak{P}$.*

  (i) *Every cycle $\xi \in \mathfrak{C}(\theta)$ either extends a subcycle of η or is forced by η (or both); if $\theta \in \mathfrak{C}$, then only one of these occurs unless $\xi = \eta$;*
  (ii) *if $\theta \in \mathfrak{S}$, then $\theta \Rightarrow \eta$.*
  (iii) *In general, $Perm(\theta)$ extends and forces $Perm(\eta)$, and every $\xi \in \mathfrak{S}(\theta)$ can be written as the disjoint union of two subpermutations $\xi = \xi_\mathcal{B} \vee \xi_\mathcal{R}$ such that $\eta \Rightarrow \xi_\mathcal{R}$ and $\xi_\mathcal{B}$ extends a subpermutation of η.*

PROOF:

Proof of (i): Apply *3.6(i)* and then *3.3* or *3.4* to a representative of ξ in the θ-adjusted map F. When $\theta \in \mathfrak{C}$, then $\eta \in \mathfrak{C}$ and every cycle extending η forces it; thus $\xi = \eta$ is the only cycle with $\eta \Rightarrow \xi$ that can extend η, by [**Ba**].

Proof of (ii): We know from *3.2(ii)* that $\eta \in \mathfrak{S}$. Let us first prove *(ii)* when $\eta \in \mathfrak{C}_k$. Then the $\mathcal{B}$-intervals $I_j = \mathcal{R}(B_{\eta^j(1)}, \mathcal{P})$ are disjoint for $j = 0, \ldots, k-1$, and $F[I_j] = I_{j+1}$ for $j = 0, \ldots, k$ and F is θ-adjusted on $\mathcal{P}$. Thus, $F^k[I_0] \subset I_0$ and there is a fixedpoint $x_0 \in I_0$ of F^k. It follows that $F^j(x_0) \in I_j$, so $\mathcal{O}(x_0)$ represents η in F; then $\theta \Rightarrow \eta$ by *1.20*. Now when $\eta \in \mathfrak{S}$, apply the above argument to each subcycle $z_0, \ldots, z_\ell = z_0$ of η to produce x with $F^j(x) \in \mathcal{R}(B_{z_j}, \mathcal{P})$ and $F^\ell(x) = x$. Then again $\mathcal{O}(x)$ represents $z_0, \ldots, z_\ell$, and the union of the $\mathcal{O}(x)$'s intersects each $\mathcal{B}$-interval in a unique point. It follows that this union represents η in F, and $\theta \Rightarrow \eta$ again.

Proof of (iii): Given a subcycle of η, we need to produce a subcycle of θ which extends it. But if $\eta(z_i) = z_{i+1}$ for $i = 0, \ldots, \ell-1$, z_j are distinct for $j = 0, \ldots, \ell-1$

and $z_\ell = z_0$, then the finite set B_{z_0} is mapped into itself by θ^ℓ, hence has a periodic orbit. That is, some $x \in B_{z_0}$ has $\theta^{\ell m}(x) = x$ for some $m \geq 1$. Since for $i \not\equiv j \pmod{\ell}$ $\theta^j(x) \neq \theta^i(x)$, we see that $per(x) = \ell m_1$ for some $m_1 \geq 1$. Thus, $\{\theta^j(x) \mid j = 0, \ldots, \ell m_1 - 1\}$ is partitioned by the sets B_{z_j}, $i = 0, \ldots, \ell - 1$ and this gives a block structure for the subcycle $\mathcal{O}(x)$ of θ with reduction $z_0, \ldots, z_{\ell-1}$. Note that $\mathcal{O}(x)$ extends and forces this subcycle, by (ii). Now, we have shown that every subcycle of η is extended and forced by some subcycle of θ, and the first part of (iii) follows.

For the second part, given $\xi \in \mathfrak{S}(\theta)$, find a representative $\mathcal{Q}$ of θ in the θ-adjusted map F such that $\mathcal{Q} \cap Flat(F, \mathcal{P}) = \emptyset$. $\mathcal{Q}$ is a union of periodic F-orbits, so by $3.6(i)$ $\mathcal{Q} = \mathcal{Q}_\mathcal{B} \cup \mathcal{Q}_\mathcal{R}$, where $\mathcal{Q}_\mathcal{B} \subset \mathcal{R}(\mathcal{B}, \mathcal{P})$ and $\mathcal{Q}_\mathcal{R}$ is contained in the $\mathcal{B}$-gaps. Now let $\xi_\mathcal{B}$ (resp. $\xi_\mathcal{R}$) be the permutation represented by $\mathcal{Q}_\mathcal{B}$ (resp. $\mathcal{Q}_\mathcal{R}$). We have $\xi = \xi_\mathcal{B} \vee \xi_\mathcal{R}$, and by 3.3, $\xi_\mathcal{B}$ extends some subpattern of η, which is a subpermutation by 3.2, while by 3.4, $\eta \Rightarrow \xi_\mathcal{R}$. ∎

4. Irreducible Patterns and Markov Graphs

In this section we relate block structures for $\theta \in \mathfrak{P}_n$ to the structure of the associated Markov graph and its transition matrix. Abstractly, the **Markov graph** of $\theta \in \mathfrak{P}_n$ is a directed graph $\mathfrak{M}(\theta)$ with vertices J_i, $i = 1, \ldots, n-1$ and an edge $J_i \to J_j$ iff j and $j+1$ lie between $\theta(i)$ and $\theta(i+1)$. A more geometric interpretation is to look at a map F which is θ-monotone on $\mathcal{P}$ (in particular, at any θ-adjusted map) and to associate the ith $\mathcal{P}$-interval (counting left-to-right) with J_i; the existence of an edge $J_i \to J_j$ corresponds to an F-covering relation. This structure is a standard device for studying cycles ([**BGMY; N; Str**]); as is well known, paths in $\mathfrak{M}(\theta)$ are in one-to-one correspondence with proper itineraries of F-orbits with respect to $\mathcal{P}$.

We will call a set Σ of vertices in $\mathfrak{M}(\theta)$ **invariant** if every edge emanating from a vertex in Σ terminates in a vertex in Σ. A graph is **strongly connected** if no proper subset of the vertices is invariant. The existence of block structures for θ is reflected in invariant sets for $\mathfrak{M}(\theta)$, as shown below.

4.1. THEOREM. *Let $\theta \in \mathfrak{P}_n$ and $\mathfrak{M}(\theta)$ its Markov graph; let F be θ-linear on $\mathcal{P} = \{1, \ldots, n\}$.*

(i) *If $\mathcal{B} = \{B_j\}_{j=1}^k$ is a block structure for θ with reduction $\eta \in \mathfrak{P}$, then the set $\Sigma_\mathcal{B}$ of vertices J_i in $\mathfrak{M}(\theta)$ for which J_i is not a $\mathcal{B}$-gap is an invariant set. Furthermore, if $\mathfrak{M}_\mathcal{R} \subset \mathfrak{M}(\theta)$ is the set of all $\mathcal{B}$-gap vertices, together with all edges in $\mathfrak{M}(\theta)$ which terminate (and hence originate) in a $\mathcal{B}$-gap vertex, then $\mathfrak{M}_\mathcal{R}$ is isomorphic, as a directed graph with an ordered set of vertices, to $\mathfrak{M}(\eta)$, the Markov graph of the reduction.*

(ii) *If $\Sigma \subset \mathfrak{M}(\theta)$ is an invariant set of vertices, there is a unique block structure $\mathcal{B}$ for which $\Sigma = \Sigma_\mathcal{B}$ as in (i).*

(iii) *Thus, the graph $\mathfrak{M}(\theta)$ is strongly connected if and only if θ has no nontrivial reductions.*

PROOF:

Proof of (i): The invariance of $\Sigma_\mathcal{B}$ follows from the observation that $\mathcal{R}(\mathcal{B}, \mathcal{P})$ is simply the union of $\mathcal{P}$-intervals corresponding to vertices in $\Sigma_\mathcal{B}$, and $\mathcal{R}(\mathcal{B}, \mathcal{P})$ is invariant.

In the discussion proving *3.4*, we noted that the quotient map $\pi : \mathcal{P} \to \mathcal{P}_\mathcal{B}$ induces an order-preserving bijection π of $\mathcal{B}$-gaps with $\mathcal{P}_\mathcal{B}$-intervals; in the present context this is an isomorphism between vertices of $\mathfrak{M}_\mathcal{R}$ and vertices of $\mathfrak{M}(\eta)$. Furthermore, we observed that a $\mathcal{B}$-gap I_i F-covers a $\mathcal{B}$-gap I_j iff $\pi(I_i)$ g-covers $\pi(I_j)$, for g any η-linear map. This is the isomorphism between edges of $\mathfrak{M}_\mathcal{R}$ and those of $\mathfrak{M}(\eta)$.

Proof of (ii): Given $\Sigma = \{I_1, \ldots, I_\ell\}$ an invariant set of vertices in $\mathfrak{M}(\theta)$ and F θ-linear on $\mathcal{P} = \{1, \ldots, n\}$, let $\mathcal{R}(\Sigma)$ be the union of $\mathcal{P}$-intervals corresponding to elements of Σ, together with any points of $\mathcal{P}$ not contained in this union. Now, let $\mathcal{R}(B_1), \ldots, \mathcal{R}(B_k)$ be the components of $\mathcal{R}(\Sigma)$ ($k > \ell$ if some points of $\mathcal{P}$ are components of $\mathcal{R}(\Sigma)$). A component is either a single point or a union of adjoining Σ-intervals. Since the F-image of every $\mathcal{P}$-interval J_i is a connected union of $\mathcal{P}$-intervals J_j corresponding to endpoints of edges $J_i \to J_j$ in $\mathfrak{M}(\theta)$, the image of each Σ-interval is contained in a single $\mathcal{R}(B_j)$, and it follows that

$F[\mathcal{R}(B_j)]$ is a subset of a single $\mathcal{R}(B_{\eta(j)})$. One sees then that $B_j = \mathcal{P} \cap \mathcal{R}(B_j)$ gives a block structure $\mathcal{B}$ for θ, with $\mathcal{R}(\mathcal{B}, \mathcal{P}) = \mathcal{R}(\Sigma)$, and $\Sigma_{\mathcal{B}} = \Sigma$.

Proof of (iii): We see immediately in *(i)* that $\Sigma_{\mathcal{B}}$ is a proper subset if and only if there is at least one $\mathcal{B}$-interval with interior and at least one $\mathcal{B}$-gap. Thus, *(iii)* is an immediate consequence of *(i)* and *(ii)*. ∎

A few observations are in order concerning our terminology. Associated to any directed graph $\mathfrak{M}$ is its **transition matrix** (or **incidence matrix**) $A = A(\mathfrak{M})$, defined by $A = [a_{ij}]$, where $a_{ij} \in \{0, 1\}$ with $a_{ij} = 1$ if and only if $J_i \to J_j$ is an edge in $\mathfrak{M}$. The transition matrix is of course non-negative, and a non-negative $n \times n$ matrix A is called **irreducible** if $A + A^2 + \cdots + A^n > 0$; that is, if for each pair i, j there exists $k \in \{1, \ldots, n\}$ for which the i, j entry of A^k is not zero. This is the same as saying that there is a path in $\mathfrak{M}$ from J_i to J_j (which can always be chosen of length at most n, if there are n vertices) for every i and j; it clearly implies the nonexistence of a proper invariant set in $\mathfrak{M}$.

There is a trivial exception to the opposite implication. If $\mathfrak{M}$ has one vertex and no edges, then there is no proper invariant set (so $\mathfrak{M}$ is strongly connected in our terminology) but the matrix $A(\mathfrak{M})$ is the 1×1 zero matrix, which is not irreducible in the (standard) sense above. There are two patterns with this exceptional Markov graph, both in $\mathfrak{P}_2$ and consisting of a single fixedpoint with one other preimage. It is easy to see that these patterns have no nontrivial block structure. Furthermore, if in a directed graph $\mathfrak{M}$ some vertex J has no edges emanating from it, then $\{J\}$ is an invariant set, and the presence of any other vertices makes this a proper invariant set. Thus, a strongly connected graph with more than one vertex must have an edge emanating from each vertex. It is easy to see that strong connectivity of a graph $\mathfrak{M}$ with no " dead end" vertices is equivalent to irreducibility of its transition matrix $A(\mathfrak{M})$.

In view of *4.1* and the preceding discussion, we adopt the following terminology.

4.2. DEFINITION. *$\theta \in \mathfrak{P}$ is an **irreducible pattern** iff there are no non-trivial reductions of θ.*

We have seen, then, that for $n > 2$, the following are equivalent:

(i) $\theta \in \mathfrak{P}_n$ is an irreducible pattern;
(ii) $\mathfrak{M} = \mathfrak{M}(\theta)$ is a strongly connected graph;
(iii) $A = A(\mathfrak{M})$ is an irreducible matrix.

The equivalence of *(i)* and *(ii)* remains true if $n = 2$; in fact all four elements of $\mathfrak{P}_2$ are irreducible patterns, and the (strongly connected) Markov graph has a single vertex, with or without an edge that begins and ends there. In this case, A is 1×1 and is either $\mathbf{I}_1 = [1]$ (irreducible) or $\mathbf{0}_1 = [0]$ (not irreducible). For $n = 1$, the unique pattern in $\mathfrak{P}_1$ is trivial and trivially irreducible; its "Markov graph" $\mathfrak{M}$ (and transition matrix A) are both empty.

In the theory of non-negative matrices [S] a somewhat stronger condition than irreducibility is **primitivity primitive matrix** (or **aperiodicity**) **aperiodic matrix** which requires that $A^k > 0$ for some k (and hence for $k = n$). In terms of the graph $\mathfrak{M}$, aperiodicity of $A(\mathfrak{M})$ means that any pair of vertices can be joined by a path of length precisely k (independent of the choice of vertices). The following shows that in our context the distinction between irreducibility and primitivity is of limited interest.

4.3. PROPOSITION. *Suppose $\theta \in \mathfrak{P}_n$ is irreducible but $A(\mathfrak{M})$ is not primitive. Then:*

(i) *There exists a unique $x \in \mathcal{P} = \{1, \ldots, n\}$ such that $\theta(x) = x$.*

(ii) For $i \in \mathcal{P}$, if $i \leq x$ then $\theta(i) \geq x$, while if $i \geq x$ then $\theta(i) \leq x$.

(iii) If F is θ-adjusted on $\mathcal{P}$, then every $\mathcal{P}$-interval contains a preimage of x under some iterate F^i.

(iv) If $\theta \in \mathfrak{S}$, either $\theta = (1) \in \mathfrak{S}_1$ or $\theta = (1\ 3)(2) \in \mathfrak{S}_3$.

Condition *(ii)* is called *division at x* in [**LMPY**] and *separation by x* in [**N**]. Note also that our description includes the exceptional patterns in $\mathfrak{P}_2$ which are irreducible but whose transition matrix A is $\mathbf{0}_1$.

PROOF: Let F be θ-adjusted on $\mathcal{P}$. The map F has a fixedpoint x (not a priori in $\mathcal{P}$). Suppose J is a $\mathcal{P}$-interval containing x. For any $j \geq 0$, $F^j[J]$ is an interval containing x. If ever $F^{j+1}[J] \supset F^j[J]$, we claim θ is aperiodic. To see this, note that then $F^j[J] \subset F^{j+1}[J] \subset \ldots$ is an increasing union of intervals and by strong connectivity, $F^{n-1}[J] = [1, n]$, so $f^j[J] = [1, n]$ for all $j \geq n$. Thus, given J' any $\mathcal{P}$-interval and $j \geq n$, there exists a path from J to J' of length precisely j. Now, for each J' there exists a path from J' to J of length $\ell = \ell(J', J)$; but note that by adjoining to this path a loop of length j from J to J we obtain a path from J' to J of length precisely $\ell + j$, for any $j \geq n$; i.e., there is a path from J' to J of length precisely k whenever $k \geq \ell(J', J) + n$. Now, take $k_0 = \max\{n + \ell(J', J) \mid J'$ a vertex in $\mathfrak{M}(\theta)\}$. Then for any J', J'' and any $k \geq k_0 + n$, there exists a path of length k from J' to J'': go from J' to J in k_0 steps, then from J to J'' in $k - k_0 \geq n$ steps. Thus, θ is aperiodic if $F^{j+1}[J] \supset F^j[J]$ for any j. Assume then that $F^{j+1}[J]$ does not contain $F^j[J]$ for any j.

Now, if x is interior to $F[J]$, then $F[J] \supset J$ contrary to our assumption. Thus x is an endpoint of $F[J]$, so $x \in \mathcal{P}$. Furthermore, if $F[J]$ contains any point of J other than x, it must contain J. Since irreducible patterns have no flat intervals, it follows that $F[J]$ is a non-degenerate interval, so $F[J]$ and J are non-degenerate intervals abutting at x. In particular, $F[J] \supset \tilde{J}$, where $\tilde{J}$ is the other $\mathcal{P}$-interval with x an endpoint. Applying the preceding argument to $\tilde{J}$, we see that $F[\tilde{J}]$ is a nondegenerate interval abutting $\tilde{J}$ at x. Thus $F^2[J] \supset F[\tilde{J}] \supset J$. Similarly, $F^2[\tilde{J}] \supset \tilde{J}$. Now suppose $F^j[J]$ and $F^{j+1}[J]$ intersect in more than one point. Then their intersection contains either J or $\tilde{J}$, and since $F[\tilde{J}] \supset J$, by increasing j by one, we get $J \subset F^j[J] \cap F^{j+1}[J]$. Take n_1 to be the least j for which this inclusion holds. We know $n_1 > 0$; note that if n_1 is odd, then $F^{n_1-1}[J] \supset J$, so n_1 is not minimal. Thus n_1 is even. This means we have paths from J to J in $\mathfrak{M}(\theta)$ of all even lengths as well as for the odd length $n_1 + 1$. We can join these together to get loops of any odd length $> n_1$. Thus there are loops $J \to J$ of length j for any $j \geq n_1$. But this allows us to apply the earlier argument to produce a path from J' to J'' of any length bigger than some given number, and $A(\mathfrak{M}(\theta))$ is aperiodic.

Thus, if A fails to be aperiodic, $F^j[J]$ abuts $F^{j+1}[J]$ at x for every j. This gives us *(ii)*.

Now, if J, $\tilde{J}$ are the only $\mathcal{P}$-intervals, we have $F[J] = \tilde{J}$, $F[\tilde{J}] = J$ and $\theta = (1\ 3)(2)$. Otherwise, let $i \neq x - 1, x$ and consider $J' = [i, i+1]$. Then J' must F^k-cover J for some k, by strong connectivity. It follows that some point of J' hits x under F^k. Since $J' \subset F^j[J]$ for some j, we cannot have x interior to $F^k[J']$. Thus either an endpoint of J' contains x in its orbit, or some interior point of J' hits a turning point or flat interval of F whose image is x. In any case, x has another pre-image in $\mathcal{P}$, and so $\theta \notin \mathfrak{S}$. ∎

A second observation concerns irreducible reductions of a pattern. Note that if θ is not irreducible, it has finitely many reductions, and we can pick a reduction η whose order is minimal among non-trivial reductions of θ. Since reduction is a transitive relation, η is irreducible. Thus, every pattern has an irreducible reduction. We ask how canonical this reduction is; an answer for permutations is given by the following.

4.4. PROPOSITION. *Suppose $\eta \neq \widetilde{\eta} \in \mathfrak{S}$ are nontrivial, irreducible, and have a common extension $\theta \in \mathfrak{S}$. Then (see figure 6)*

$$\eta = (1\ 3)(2), \quad \widetilde{\eta} = (1\ 2), \quad (\text{or vice versa})$$

and θ extends $\theta_0 = (1\ 4)(2\ 3)$.

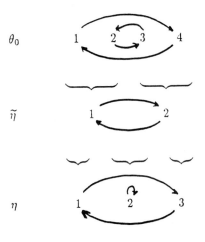

Figure 6

PROOF: Suppose $\mathcal{B}$, $\widetilde{\mathcal{B}}$ are block structures for θ over η, $\widetilde{\eta}$ respectively. Consider the common refinement $\mathcal{B} \vee \widetilde{\mathcal{B}}$. This is automatically a block structure for θ over some $\theta' \in \mathfrak{S}$, and each of $\mathcal{B}$, $\widetilde{\mathcal{B}}$ induces a block structure for θ' over η, $\widetilde{\eta}$ respectively. Thus, we can replace θ with θ' until $\theta = \theta'$. This means $\mathcal{B} \vee \widetilde{\mathcal{B}}$ consists of individual points of $\mathcal{P} = \{1, \ldots, n\}$.

Now consider the $\mathcal{B}$-gaps. Any $\mathcal{P}$-interval which F-covers a $\mathcal{B}$-gap is itself a $\mathcal{B}$-gap. Thus, if some $\mathcal{P}$-interval J is both a $\mathcal{B}$-gap and a $\widetilde{\mathcal{B}}$-gap, consider the collection of all $\mathcal{P}$-intervals which F^j-cover J for some j. Then these are all simultaneously $\mathcal{B}$-gaps and $\widetilde{\mathcal{B}}$-gaps, and their complement gives a block structure $\mathcal{B}^*$ of which $\mathcal{B}$ and $\widetilde{\mathcal{B}}$ are both refinements. But then $\mathcal{B}^*$ induces a non-trivial block structure for η and $\widetilde{\eta}$, contrary to their irreducibility. Thus, every $\mathcal{B}$-gap is contained in a $\widetilde{\mathcal{B}}$-interval, and every $\widetilde{\mathcal{B}}$-gap is contained in a $\mathcal{B}$-interval. Let us call a $\mathcal{B}$-interval (resp. $\widetilde{\mathcal{B}}$-interval) **nontrivial** if it contains more than one point of $\mathcal{P}$. Let Σ consist of all points in the intersection of nontrivial $\mathcal{B}$-intervals with nontrivial $\widetilde{\mathcal{B}}$-intervals. Note that, since θ is a permutation, the F-image of a nontrivial $\mathcal{B}$ (or $\widetilde{\mathcal{B}}$)-interval is itself nontrivial. Hence Σ is F-invariant. Also, we have assumed that *(i)* $\mathcal{B}$ and $\widetilde{\mathcal{B}}$ have no common gaps, and *(ii)* the intersection of any $\mathcal{B}$-interval with any $\widetilde{\mathcal{B}}$-interval is empty or trivial.

Now, consider the leftmost element $1 \in \mathcal{P}$. It belongs to some $\mathcal{B}$-interval and to some $\widetilde{\mathcal{B}}$-interval. If both are trivial, then $[1, 2]$ is a common gap of $\mathcal{B}$ and $\widetilde{\mathcal{B}}$. Hence one of these is nontrivial; suppose $B_1 = [1, b_1]$ is a nontrivial $\mathcal{B}$-interval. Now look at $[b_1, b_1 + 1]$. Since it is a $\mathcal{B}$-gap, it must be contained in a $\widetilde{\mathcal{B}}$-interval, $B_2 = [a_2, b_2]$, where $a_2 = b_1$ since $B_1 \cap B_2$ must be trivial.

Now consider $[b_2, b_2 + 1]$; it must belong to a nontrivial $\mathcal{B}$-interval $B_3 = [a_3, b_3]$ with $a_3 = b_2$. Inductively, we write $[1, n]$ as a union $B_1 \cup B_2 \cup B_3 \cup \ldots B_k$ of nontrivial intervals, each a $\mathcal{B}$-interval or a $\widetilde{\mathcal{B}}$-interval, with $B_i \cap B_{i+1}$ a single point in Σ. In fact, Σ is precisely the set of common endpoints of $B_i \cap B_{i+1}$, $i = 1, \ldots, k - 1$. Furthermore, there can be no other nontrivial $\mathcal{B}$(resp. $\widetilde{\mathcal{B}}$)-intervals, since such an interval would be contained in a $\widetilde{\mathcal{B}}$(resp. $\mathcal{B}$)-interval, contrary to the nontriviality of intersection. Thus, each B_i maps onto another B_j. Now, note that if any B_i has more than one interior point in $\mathcal{P}$, then so do its images. If B_i is a $\mathcal{B}$-interval, these interior blocks are unions of (trivial) $\widetilde{\mathcal{B}}$-intervals and give a non-trivial reduction of $\widetilde{\eta}$, contrary to its irreducibility. Thus, each nontrivial interval has either one or no interior point in $\mathcal{P}$.

Now, consider the configuration. We have a set $\Sigma \subset \mathcal{P}$, invariant under θ. Furthermore, between two successive elements of Σ is at most one point of $\mathcal{P}$, and the block of 2 or 3 points formed by these is mapped precisely to another such block. It follows that the map θ is strictly monotone on $\mathcal{P}$: thus either $\theta = id$ or θ is a flip.

If $\theta = id$, then η and $\widetilde{\eta}$ are both the identity on some set. The identity on a single point is trivial, and the identity on a set of at least three points is reducible. Thus, in this case, η and $\widetilde{\eta}$ must both be the identity on two points. This contradicts the hypothesis that $\eta \neq \widetilde{\eta}$.

On the other hand, of θ is a flip, so are both η and $\widetilde{\eta}$. The reductions of a flip of odd degree are flips of any lower odd degree. Thus, if θ is a flip on an odd number of elements, then η and $\widetilde{\eta}$ must both be flips of degree 3, contradicting $\eta \neq \widetilde{\eta}$. On the other hand, a flip of even degree (at least 4) has two irreducible reductions, namely the flips of degree 2 and degree 3. Thus, if θ is a flip of even degree, and has two distinct irreducible reductions, these must be flips of respective degree 2 and 3, and θ must extend the flip of degree 4, $\theta_0 = (1\ 4)(2\ 3)$.

This completes the proof. ∎

Accordingly, we can say that there is a "canonical" reduction of any permutation θ. If θ does not extend $\theta_0 = (1\ 4)(2\ 3)$, it has a *unique* irreducible reduction (which may be one of $(1\ 3)(2)$ or $(1\ 2)$), whereas if it extends $(1\ 4)(2\ 3)$, this is the canonical reduction, capturing the fact that it reduces further in precisely two ways.

5. Horseshoe Patterns and Fold Type

Various notions of "horseshoes" for maps of the interval have been used in calculating entropy ([**MS;M;Bl;BGMY**]) and characterizing forcing-minimal cycles ([**BlC;C**]). Also, the classification of cycles according to the number and character of their turning points–and especially the class of *unimodal* cycles–has played a role in many studies in this area ([**Ba;Ca;CE;J;MT**]). In this section, we formalize both notions in terms of patterns and study some aspects of the role of each in the forcing relation. Our results will be used in §§6,8,9, and 11.

5.1. DEFINITION. *Given $k \geq 2$ and $\sigma \in \{\pm 1\}$, the k-horseshoe with sign σ is the pattern (see figure 7)*

$$\mathfrak{h}(k, \sigma) \in \mathfrak{P}_{k+1}$$

defined by

$$\mathfrak{h}(k, \sigma)(i) = \begin{cases} 1 & \text{if } (-1)^{i+1} = \sigma \\ k+1 & \text{if } (-1)^{i+1} = -\sigma \end{cases}$$

for $i = 1, \ldots, k+1$.

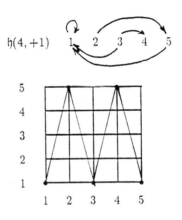

$$\mathfrak{h}(4, +1)$$

Figure 7

A useful device for detecting k-horseshoes is the following:

5.2. LEMMA. *A continuous map $f \in \mathcal{E}(I)$ exhibits $\mathfrak{h}(k, \sigma)$ if and only if there exist k intervals $I_1 \leq I_2 \cdots \leq I_k$ with disjoint interiors contained in I such that each I_i f-covers all the I_j's, and I_1 f-covers each I_j with index σ.*

PROOF: If f exhibits $\mathfrak{h}(k, \sigma)$ on $\mathcal{P} = \{x_1, \ldots, x_{k+1}\}$ then clearly $I_j = [x_j, x_{j+1}]$ have the required property.

Conversely, suppose $I_j = [x_j, y_j]$ are subintervals of I which f-cover $[x_1, y_k]$, and in particular I_1 f-covers $[x_1, y_k]$ with index σ. Cutting down I if necessary, we can assume each I_j f-covers I (I_1 doing so with index σ). Now, cutting down each I_j if necessary, we can assume further that each I_j minimally f-covers I. Also, if $f(y_j) \neq f(x_{j+1})$ for some $j \in \{1, \ldots, k-1\}$, then $[y_j, x_{j+1}]$ also f-covers I, and we can replace I_{j+1} with a subinterval $\widetilde{I}_j = [\widetilde{x}_{j+1}, \widetilde{y}_{j+1}]$ of $[y_j, x_{j+1}]$ which minimally f-covers I so that $f(y_j) = f(\tilde{x}_{j+1})$. Thus, by induction on j, we can assume for each $j = 1, \ldots, k - 1$ that

$$f(x_j) \neq f(y_j) = f(x_{j+1}) \neq f(y_{j+1}).$$

Since all these values are either a or b, we have $f(x_j) = f(y_{j+1})$. It follows that the index of f on I_j alternates with j, starting at σ on I_1: that is,

I_j f-covers $I = [a, b]$ (minimally) with index $\sigma_j = (-1)^{j+1}\sigma$ for $j = 1, \ldots, k$.

Now, apply *1.1(i)* to each $I_j = I$ with $J = [x_1, y_k]$ and the finite set $\{x_1, y_k\}$ to find points $u_j, v_j \in I_j$ with $u_j <_{\sigma_j} v_j$ and $f(u_j) = x_1$, $f(v_j) = y_k$. Note that we can take $u_1 = x_1$ and $v_k = y_k$. Then we have

$$f(u_j) = f(v_{j+1}) = \begin{cases} x_1 & \text{if } \sigma_j = +1 \\ y_k & \text{if } \sigma_j = -1. \end{cases}$$

Now let $\mathcal{Q} = \{u_1, \ldots, u_k, v_k\}$. Then

$$f(u_j) = \begin{cases} x_1 = u_1 & \text{if } \sigma = (-1)^{j+1} \\ y_k = v_k & \text{if } -\sigma = (-1)^{j+1} \end{cases}$$

and

$$f(v_j) = \begin{cases} x_1 & \text{if } \sigma = (-1)^{k+1} \\ y_k & \text{if } -\sigma = (-1)^{k+1} \end{cases}$$

which clearly shows that f exhibits $\mathfrak{h}(k, \sigma)$ on $\mathcal{Q}$. ∎

Many studies of maps of the interval concentrate on *unimodal* maps–those with a unique turning point. The combinatorics of periodic orbits for such maps is simpler than the general case, and is handled especially effectively by the kneading calculcus ([**MT;CE**]), which shows in particular that forcing linearly orders such cycles. The role of "modality" in the forcing relation when more than one turning point is present has been considered by Baldwin[**Ba**] and Jungreis[**J**].

Here, we define the *fold type* of a pattern, which measures the "modality" by means of the *lap number* of [**MT**].

5.3. DEFINITION. *Given $\theta \in \mathfrak{P}_n$, define the **lap number** $k \in \{0, 1, 2, \ldots\}$ of θ as follows:*

(i) *If θ is constant (i.e., $\theta(i) = \theta(1)$ for all i), then $k = 0$.*

(ii) *If θ is not constant, it is possible to pick*

$$1 = i_0 < i_1 < \cdots < i_k = n$$

and $\sigma \in \{\pm 1\}$ such that for $j = 0, \ldots, k - 1$,

$$\theta(i_j) \neq \theta(i_{j+1})$$

and

$$\theta(i) \leq_{(-1)^j \sigma} \theta(i + 1)$$

whenever $i_j \leq i < i_{j+1}$.

The **fold type** of θ is 0 in case *(i)* and (k, σ) in case *(ii)*. We refer to θ as a *k*-**fold pattern** *(or, if $k > 0$ and σ matters, as a (k, σ)-**fold pattern**).*

It is easy to see that the numbers i_j constitute one point from each critical element of θ. A maximal block (resp. interval) on which a pattern (resp. map) is weakly monotone is sometimes called a **lap**. The lap number k of θ, if nonzero, counts the number of laps for θ; the number of critical elements for θ is $k + 1$, while the number of turning points is $k - 1$. In particular, a **unimodal pattern** is a k-fold pattern with $k \leq 2$.

It follows easily from *2.1(i)* and *5.3* that a k-fold pattern cannot force any patterns with higher lap number. One might, however, expect that the horseshoe pattern $\mathfrak{h}(k, \sigma)$, which has fold type (k, σ), would force any other pattern of the same fold type. But a $\mathfrak{h}(k, \sigma)$-adjusted map F is piecewise expanding, so that any pair of tandem cycles in a pattern exhibited by F must be separated by a turning point of F. Similarly, *2.1(ii)* implies that the points of any flat block for such a pattern must also be separated by turning points of F. These requirements prevent $\mathfrak{h}(k, \sigma)$ from forcing certain patterns of the same fold type. However, as we see below, in the absence of these phenomena our expectation is met.

5.4. PROPOSITION. *Suppose $\theta, \eta \in \mathfrak{P}$, with θ m-fold and η k-fold.*

(i) *If $\theta \Rightarrow \eta$, then either $k = m$ and they have the same fold type, or $m > k$.*
(ii) *Suppose $\theta = \mathfrak{h}(m, \sigma)$ and η satisfies*
   (a) either $m = k$ and η is (m, σ)-fold, or $m > k$;
   (b) η has no flat blocks;
   (c) every tandem block of η contains at most three points, and both end-points of the block belong to a single cycle which contains a critical point of η;
   then $\theta \Rightarrow \eta$.

Note that our hypotheses in *(ii)* are automatically satisfied if η is either a cycle or a k-horseshoe with $k < m$.

PROOF:

Proof of (i): This is automatic when $m = 0$.

If $m > 0$, let F be θ-adjusted on $\mathcal{P}$ and take $\mathcal{Q}$ a representative of η in F with $\mathcal{Q} \cap Flat(F, \mathcal{P}) = \emptyset$. Let $y_0 < y_1 < \cdots < y_k$ be the points in $\mathcal{Q}$ corresponding to the integers $i_0, i_1, \ldots, i_k$ in definition *5.3*. The points $x_0 < x_1 < \cdots < x_k$ in $\mathcal{P}$ yielded by *2.1* belong to distinct turning elements of θ; thus $m \geq k$. Furthermore, if $m = k$ then there are no other turning elements of θ, and the fact that x_1 and y_1 are both σ-maximal completes the proof of *(i)*.

Proof of (ii): Note first that the three trivial cases, $\eta = (1)$, $\eta = (1\ 2)$, and $\eta = (1\ 3)(2)$, are all forced by every horseshoe. In every other case for which *(a)-(c)* hold, we have $k \geq 1$.

Note also that *(ii)* holds if η is a horseshoe. To see this, suppose we have a map F which is θ-adjusted on $\{1, \ldots, m\}$, and consider the collection of k intervals $\{[i, i + 1]\}_{i=a}^{a+k-1}$ where $a = 1$ if $\widetilde{\sigma} = \sigma$ and $a = 2$ if $\widetilde{\sigma} = -\sigma$: since $k < m$, we have $a + k \leq m$, so these intervals satisfy the hypotheses of *5.2*, thus giving us a representative of $\mathfrak{h}(k, \widetilde{\sigma})$ in F.

It remains to show that $\mathfrak{h}(k, \sigma)$ forces any (k, σ)-fold pattern η which satisfies *(b)* and *(c)*. Let $1 = t_0 < t_1 < \cdots < t_k = n$ be the critical points of $\eta \in \mathfrak{P}_n$ (by *(b)*, $t_j = i_j$ in *5.3*). Form $\widetilde{\mathcal{P}}$ by adjoining to $\mathcal{P} = \{1, \ldots, n\}$ a collection of $k + 1$ points t_i^*, defined as follows:

(1) if 1 is a fixedpoint or $\{1, n\}$ is a 2-cycle in η, let $t_0^* = 1$; otherwise $t_0^* = 0$;

(2) if n is a fixedpoint or $\{1, n\}$ is a 2-cycle in η, let $t_k^* = n$; otherwise $t_k^* = n + 1$;

(3) for $j = 1, \ldots, k-1$, $t_j^* = t_j \pm \frac{1}{3}$, with the choice of sign limited only by the requirement that if t_j belongs to a tandem block of η, then t_j^* is interior to that block.

Note that if a tandem block of η contains t_0 or t_k, then (assuming η is not one of the trivial cases above) it is the image under η or η^2 of another t_j, $1 \le j \le k-1$. Thus by (c), every tandem block of η encloses some t_j^*, $1 \le j \le k - 1$. Now define $f : \widetilde{\mathcal{P}} \to \widetilde{\mathcal{P}}$ by

$$f(t_j) = \eta(t_j)$$
$$f(t_j^*) = \begin{cases} t_k^* & \text{if } (-1)^j = \sigma \\ t_0^* & \text{if } (-1)^j = -\sigma. \end{cases}$$

The reader can easily check that the pattern $\widetilde{\eta}$ exhibited by f on $\widetilde{\mathcal{P}}$ has the t_j^*'s as its turning points, and has no flat or tandem blocks. Thus by 2.6 $\widetilde{\eta}_{**}$ is exhibited by the restriction of f to the t_j^*'s, and it is clear that this is just $\mathfrak{h}(m, \sigma)$, so $\mathfrak{h}(m, \sigma) \Rightarrow \widetilde{\eta}$. But η is a subpattern of $\widetilde{\eta}$, and it follows that $\mathfrak{h}(m, \sigma) \Rightarrow \eta$. ∎

In §11 we will make use of the possibility of finding a k-fold cycle which forces a $(k-2)$-horseshoe. This is a harder problem than that treated in 5.4(ii): if $\theta \Rightarrow \mathfrak{h}(k, \sigma)$ and θ is a permutation, the fold type of θ is more restricted than 5.4 might lead us to expect. The situation is described in the following.

5.5. PROPOSITION. *Given* $2 \le \widetilde{k} < k$ *and* $\sigma, \widetilde{\sigma} \in \{\pm 1\}$,

(i) *if there exists* $\theta \in \mathfrak{S}$ *of fold type* (k, σ) *with* $\theta \Rightarrow \mathfrak{h}(\widetilde{k}, \widetilde{\sigma})$, *then we can choose* $\theta \in \mathfrak{C}$;

(ii) *the only cases when no such* θ *exists are*
   (a) $\widetilde{k} = k - 1$ *and* $\sigma = \widetilde{\sigma} = +1$;
   (b) $\widetilde{k} = k - 2$, k *is odd and* $\sigma = \widetilde{\sigma} = +1$.

This result will follow from some special cases, given in lemmas 5.6-5.8.

5.6. LEMMA. *For* $k \ge 4$ *and* $\sigma = \pm 1$, *there exists a cycle* $\theta \in \mathfrak{C}_{4(k-2)}$ *such that*

$$\mathfrak{h}(k, \sigma) \Rightarrow \theta \Rightarrow \mathfrak{h}(k - 2, -\sigma)$$

PROOF: Let F be $\mathfrak{h}(k, \sigma)$-adjusted on $\mathcal{P} = \{1, \ldots, k+1\}$. Note that every $\mathcal{P}$-loop is proper. Consider the loop $\mathcal{I} = \{J_i\}_{i=0}^{4(k-2)}$ defined by

$$J_{2j} = \begin{cases} [j+2, j+3] & j = 0, \ldots, k-3 \\ [j+4-k, j+5-k] & j = k-2, \ldots, 2k-5 \end{cases}$$

$$J_{2j+1} = \begin{cases} [1, 2] & j = 0, \ldots, k-3 \\ [k, k+1] & j = k-2, \ldots, 2k-5 \end{cases}$$

$$J_{4(k-2)}(= J_0) = [2, 3].$$

That is, each $\mathcal{P}$-interval $I_j = [j, j+1]$ other than $I_1 = [1, 2]$ and $I_k = [k, k+1]$ occurs twice among $J_0, \ldots, J_{4k-9}$, the first time followed by I_1 and the second time by I_k. The reader can check, for example, that when $k = 4$ this scheme yields (see figure 8)

$$\mathcal{I} = \{I_2, I_1, I_3, I_1, I_2, I_4, I_3, I_4, I_2\}$$

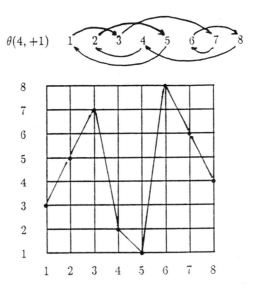

Figure 8

and

$$\theta(4, +1) = (1\ 3\ 7\ 6\ 8\ 4\ 2\ 5).$$

Since $\mathcal{I}$ is a prime loop, there is a (unique) periodic orbit $\mathcal{O}$ of least period $4(k-2)$ in the representative set $\mathcal{R}(\mathcal{I}, F)$. Let θ be the cycle represented by $\mathcal{O}$. Note that $\mathcal{O}$ has $k-2$ points each in I_1 and I_k, and two points in each I_j, $j = 2, \ldots, k-1$. Of these two, one maps into I_1 and the other into I_k under F. Thus, if G is θ-adjusted on $\mathcal{Q} = \{1, \ldots, 4(k-2)\}$, then each of the $\mathcal{Q}$-intervals $\tilde{I}_j = [k-2+2j, k-2+2j+1]$ $(j = 2, \ldots, k-1)$ G-covers $[k-2, 3(k-1)]$, which is the convex hull of $\bigcup_{j=2}^{k-1} \tilde{I}_j$. It follows that G exhibits a $(k-2)$-horseshoe. Furthermore, since the index of F on $[2,3]$ is $-\sigma$, this horseshoe exhibited by G is $\mathfrak{h}(k-2, -\sigma)$. ∎

5.7. LEMMA. *For $k \geq 3$ there exists a cycle $\theta \in \mathfrak{C}_{5k-7}$ such that*

$$\mathfrak{h}(k, -1) \Rightarrow \theta \Rightarrow \mathfrak{h}(k-1, -1)$$

PROOF: We carry out a construction similar to that in 5.6: let F be $\mathfrak{h}(k, -1)$-adjusted on $\mathcal{P} = \{1, \ldots, k+1\}$; we will exhibit θ by defining a loop $\mathcal{I} = \{J_\ell\}_{\ell=0}^{5k-7}$. The main definition breaks into two parts.

(a) For $\ell = 0, \ldots, 3(k-1) - 1 = 3k - 4$, write $\ell = 3(j-1) + \alpha$, where $j \in \{1, \ldots, k-1\}$ and $\alpha \in \{0, 1, 2\}$; then

$$J_\ell = \begin{cases} [j, j+1] & \text{if } \alpha = 0 \\ [1, 2] & \text{if } \alpha = 1 \\ [k, k+1] & \text{if } \alpha = 2 \end{cases}.$$

(b) For $\ell = 3k-5, \ldots, 5k-8 = 3k-4+2(k-2)$, write $\ell = 3k-4+2(j-2)+\beta$, where $j \in \{2, \ldots, k-1\}$ and $\beta \in \{1, 2\}$; then

$$J_\ell = \begin{cases} [j, j+1] & \text{if } \beta = 1 \\ [k, k+1] & \text{if } \beta = 2 \end{cases}.$$

Finally, we make $\mathcal{I}$ a loop by setting $J_{5k-7} = J_0 = [1, 2]$. The itinerary $\mathcal{I}$ has the form

$$\mathcal{I} = \{I_1, I_1, I_k; I_2, I_1, I_k; I_3, I_1, I_k; \ldots$$
$$\ldots; I_{k-1}, I_1, I_k; I_2, I_k; I_3, I_k; \ldots$$
$$\ldots; I_{k-1}, I_k; I_1\}$$

where $I_j = [j, j+1]$. For example, $k = 3$ yields (figure 9)

$$\mathcal{I} = \{I_1, I_1, I_3; I_2, I_1, I_3; I_2, I_3; I_1\}$$

and

$$\theta(3, -1) = (1\ 7\ 4\ 2\ 6\ 5\ 8\ 3).$$

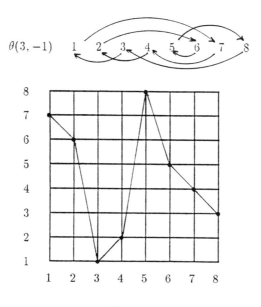

Figure 9

Now let $\mathcal{O}$ be the $(5k - 7)$-periodic F-orbit in $\mathcal{R}(\mathcal{I}, F)$ and θ the cycle it represents. The only occurrence in $\mathcal{I}$ of I_1 not followed by I_k is the first. Thus, every point of $I_1 \cap \mathcal{O}$ but one maps under F into I_k; the one exception maps to another point of I_1. Since F reverses orientation on I_1, this means the exceptional point is the rightmost point of $I_1 \cap \mathcal{O}$. The $\mathcal{O}$-interval $J_1 \subset I_1$ with this as its right endpoint thus F-covers itself and $I_2 \cup \cdots \cup I_{k-1}$, with index -1. Now, for $j = 2, \ldots, k - 1$, I_j appears twice in $\mathcal{I}$, once followed by I_k and once by I_1, I_k. The corresponding points of $\mathcal{O}$ map, respectively, to I_k and to a point of $I_1 \cap \mathcal{O}$ other than the exceptional point. Thus, the unique $\mathcal{O}$-interval J_j in I_j $(j = 2, \ldots, k - 1)$ F-covers $J_1 \cup J_2 \cup \cdots \cup J_{k-1}$. It follows that θ forces $\mathfrak{h}(k - 1, -1)$. ∎

5.8. Proposition. *A permutation of fold type $(k, +1)$ cannot force $\mathfrak{h}(m, +1)$ if*

(a) $m = k - 1 \geq 2$, *or*

(b) $m = k - 2 \geq 3$ *is odd.*

PROOF: Let $\theta \in \mathfrak{S}_n$ be a $(k, +1)$-fold permutation, and F θ-adjusted on $\mathcal{P} = \{1, \ldots, n\}$. Let $z \in \mathcal{P}$ be the right endpoint of the first lap of F: thus, $1 \le x < y \le z$ implies $F(x) \le F(y)$, while $F(z) > F(z+1)$. Note that since F is nondecreasing on $[1, z]$ and θ is a permutation, we have $\theta(x) \ge x$ for $x \in \mathcal{P}$ in the first lap of θ. Therefore, since F is θ-adjusted on $\mathcal{P}$, every fixedpoint x of F in $[1, z]$ must belong to $\mathcal{P}$. If k is odd, then F is also nondecreasing on its last lap, $[z', n]$, and this also says every fixedpoint x of F in $[z', n]$ belongs to $\mathcal{P}$.

Let $\mathfrak{h} = \mathfrak{h}(m, +1)$, where $m = k - 1$ or $m = k - 2$, and suppose $\mathcal{Q}$ is a representative of $\mathfrak{h}$ in F. Note that the leftmost point of $\mathcal{Q}$ (as well as the rightmost if m is odd) is fixed. Thus, if $\mathcal{Q} \cap \mathcal{P} = \emptyset$, we must have $\mathcal{Q} \subset [z, n]$ ($\mathcal{Q} \subset [z, z']$ if m is odd). By 5.4, F must have at least $m - 1$ turning elements interior to $[z, n]$ (resp. $[z, z']$ if m is odd) with the leftmost one of minimum type (resp. and the rightmost of maximum type). But since z and z' are turning points for F with z maximum (and z' minimum, if m is odd), we have at most $k - 2$ such points in (z, n) (and $k - 3$ in (z, z')). This proves that θ cannot *strongly* force $\mathfrak{h}$ in case *(a)* and *(b)*.

Furthermore, if $F(x) = x < z$ (resp. $F(y) = y > z'$), then the intervals $[1, x]$ and $[x, n]$ (resp. $[1, y]$ and $[y, n]$) are F-invariant, and since $F|\mathcal{P}$ is one-to-one, no interior point of either can map to x (resp. y). Thus, again we must have $\mathcal{Q} \subset [z, n]$ (resp. $\mathcal{Q} \subset [z, z']$), leading to the same contradiction as above, and proving the proposition. ∎

5.9. REMARK. *(a) 5.8 holds with "permutation" replaced by "pattern" and "force" replaced by "strongly force", provided the pattern is also assumed to satisfy*

(i) *$\theta(x) \ge x$ for $x \in \mathcal{P}$ in the first lap of θ; and*
(ii) *if k is odd, $\theta(x) \le x$ for $x \in \mathcal{P}$ in the last lap of θ.*

(b) The necessity of hypotheses (i),(ii) in this more general setting is illustrated by the pattern $\theta \in \mathfrak{P}_5$ of type $(3, +1)$ given by

$$\theta(1) = \theta(2) = \theta(4) = 1$$
$$\theta(3) = \theta(5) = 5$$

which strongly forces a horseshoe of type $(2, +1)$.

The proof of *5.9(a)* is, with some obvious minor modifications, the same as the first two paragraphs of the preceding proof. We thank the referee for pointing out to us the possibility of an example like *5.9(b)*.

PROOF OF 5.5: First observe the following about the effect of flips. Given $\theta \in \mathfrak{P}_n$, define $\theta^f \in \mathfrak{P}_n$ to be the conjugate of θ by a flip:

$$\theta^f(i) = n + 1 - \theta(n + 1 - i) \quad \text{for } i = 1, \ldots, n.$$

Then

(*) $$\theta \Rightarrow \eta \text{ iff } \theta^f \Rightarrow \eta^f,$$

and

(**) $$\text{if } \theta \text{ is } (k, \sigma)\text{-fold, then } \theta^f \text{ is } (k, (-1)^{k-1}\sigma)\text{-fold.}$$

To see (**), note that the number of turning points $(k - 1)$ is invariant under conjugacy, and the orientation of θ^f to the left of its first turning point agrees with the orientation to the right of its last turning point. In particular,

$$\mathfrak{h}(k, \sigma^f) = \begin{cases} \mathfrak{h}(k, \sigma) & \text{if } k \text{ is odd} \\ \mathfrak{h}(k, -\sigma) & \text{if } k \text{ is even.} \end{cases}$$

To prove 5.5, we separate three cases.

Case 1: $\widetilde{k} = k - 1$: When $\sigma = \widetilde{\sigma} = -1$, 5.7 gives $\theta = \theta(k, -1)$ which is $(k, -1)$-fold forcing $\mathfrak{h}(k-1, -1)$. If $\sigma \neq \widetilde{\sigma}$, consider $\theta(k, -1)^f \Rightarrow \mathfrak{h}(k-1, -1)^f$; when k is even (so $k - 1$ is odd) $\theta(k, -1)^f$ is $(k, +1)$-fold and $\mathfrak{h}(k-1, -1)^f = \mathfrak{h}(k-1, -1)$, while when k is odd, $\theta(k, -1)^f$ is $(k, -1)$-fold and $\mathfrak{h}(k-1, -1)^f = \mathfrak{h}(k-1, +1)$. Finally, when $\sigma = \widetilde{\sigma} = +1$, the first case of 5.8 tells us no appropriate θ exists.

Case 2: $\widetilde{k} = k-2$: 5.6 gives us θ which is (k, σ)-fold forcing $\mathfrak{h}(k-2, \widetilde{\sigma})$ if $\widetilde{\sigma} \neq \sigma = \pm 1$. If $\sigma = \widetilde{\sigma} = -1$, 5.7 gives us θ which is (k, σ)-fold forcing $\mathfrak{h}(k-1, -1)$, which in turn forces $\mathfrak{h}(k-2, -1)$. Thus we have θ which is (k, σ)-fold forcing $\mathfrak{h}(k-2, \widetilde{\sigma})$ unless $\sigma = \widetilde{\sigma} = +1$. If k is even, take θ as in 5.7 $(k, -1)$-fold forcing $\mathfrak{h}(k-2, -1)$, then note that θ^f is $(k, +1)$-fold and forces $\mathfrak{h}(k-2, -1)^f = \mathfrak{h}(k-2, +1)$. Finally, when k is odd and $\sigma = \widetilde{\sigma} = +1$, 5.8 tells us no (k, σ)-fold $\theta \in \mathfrak{S}$ forces $\mathfrak{h}(k-2, \widetilde{\sigma})$.

Case 3: $\widetilde{k} < k - 2$: Given k, σ, by case 2 there exists some σ_1 and $\theta \in \mathfrak{C}$ which is (k, σ)-fold with $\theta \Rightarrow \mathfrak{h}(k-2, \sigma_1)$. But by 5.4, $\mathfrak{h}(k-2, \sigma_1) \Rightarrow \mathfrak{h}(\widetilde{k}, \widetilde{\sigma})$ for any $\sigma_1, \widetilde{\sigma} \in \{\pm 1\}$, so $\theta \Rightarrow \mathfrak{h}(\widetilde{k}, \widetilde{\sigma})$. ∎

6. EXTENSIONS OF CYCLES

Block structures whose associated reduction is a cycle have some special features which we shall exploit in §9.

Suppose $\theta \in \mathfrak{P}$ is a pattern with block structure $\mathcal{B} = \{B_j\}_{j=1}^k$ over a cycle $\eta \in \mathfrak{C}_k$. The map θ^k takes each block $B_j = \{i + i_j \mid i = 1, \ldots, m_j\}$ into itself, and $\theta^k | B_j$ represents a **first return pattern** $\alpha_j \in \mathfrak{P}_{m_j}$ for B_j, defined by

$$\theta^k(i + i_j) = \alpha_j(i) + i_j \quad i = 1, \ldots, m_j.$$

We call θ an **extension of η by** α_j in this case. When θ is a permutation, then by *3.2(ii)* the number of elements in each block is the same: $m_j = m$ and $i_j = (j - 1)m$ for $j = 1, \ldots, k$. It is then easy to see that the first-return permutations α_j for different blocks B_j are conjugate elements of the symmetric group $\mathfrak{S}_m$. However, when η is not surjective, we have no such relation among the various patterns α_j; in fact, they will in general be defined on different numbers of elements.

We shall call $\theta \in \mathfrak{P}$ a **simple extension** of $\eta \in \mathfrak{C}$ if θ has a block structure over η such that θ fails to be (strictly) monotone on at most one block: that is, if there exists j_0 such that for each $j \neq j_0$, the map $\theta : B_j \rightarrow B_{\eta_j}$ either preserves or reverses order. Simple extensions play a central role in the characterization of forcing-minimal (or *primary*) cycles for maps of the interval and of other one-dimensional spaces ([**ALS; C; BlC; H; ALM; BCJM**]). The observation *(ii)* below was noted in [**ALM**] for $\theta \in \mathfrak{C}$.

6.1. REMARK. *Suppose $\theta \in \mathfrak{P}_n$ is a simple extension of $\eta \in \mathfrak{C}_k$, with block structure $\mathcal{B} = \{B_j\}_{j=1}^k$ such that $\theta | B_j$ is strictly monotone for $j \neq j_0$. Let α_j be the first-return on B_j.*

   (i) *if F is θ-adjusted on $\mathcal{P} = \{1, \ldots, n\}$, then for $j = 1, \ldots, k$, $F^k | \mathcal{R}(B_j, \mathcal{P})$ is α_j-adjusted on $\mathcal{P} \cap \mathcal{R}(B_j, \mathcal{P})$;*

   (ii) *if $\theta \in \mathfrak{S}$, then for any two blocks B_i, B_j the first-return patterns α_i, α_j are equal or conjugate by a flip.*

PROOF:

Proof of (i): Let $F_j = F | \mathcal{R}(B_{\eta(j)}, \mathcal{P})$. Then

$$F_j : \mathcal{R}(B_j, \mathcal{P}) \rightarrow \mathcal{R}(B_{\eta(j)}, \mathcal{P})$$

is piecewise-monotone, preserves $\mathcal{P}$, and for $j \neq j_0$ is a homeomorphism. Clearly,

$$F^k | \mathcal{R}(B_j, \mathcal{P}) = F_{\eta^{k-1}(j)} \circ F_{\eta^{k-2}(j)} \circ \cdots \circ F_j.$$

Thus, $F^k | \mathcal{R}(B_j, \mathcal{P})$ has the form $h_1 \circ F_{j_0} \circ h_2$, where h_1 and h_2 are homeomorphisms that map points of $\mathcal{P}$ to points of $\mathcal{P}$. It follows that $F^k | \mathcal{R}(B_j, \mathcal{P})$ is B_j-monotone and $F^k(i) = \theta^k(i) = \alpha_j(i)$ for $i \in B_j$; the conditions of *1.5* for F^k follow immediately from the corresponding conditions for F.

Proof of (ii): This is immediate from the representation

$$F^k | \mathcal{R}(B_j, \mathcal{P}) = h_1 \circ F_{j_0} \circ h_2. \quad \blacksquare$$

We note that the converse of *6.1(ii)* is false. For example, each of the 9-cycles

$$\theta_1 = (1\ 6\ 9\ 3\ 5\ 8\ 2\ 4\ 7)$$
$$\theta_2 = (1\ 5\ 9\ 3\ 4\ 8\ 2\ 6\ 7)$$

(see figure 10) is a 3-extension of $\eta = (1\ 2\ 3)$ with all first-return cycles α_j $(j = 1, 2, 3)$ equal to $\alpha = (1\ 3\ 2)$. The first 9-cycle, θ_1, preserves order on B_2 and reverses it on B_3, so θ_1 is a simple extension of η by α (with $j_0 = j$). The second 9-cycle, θ_2, is monotone only on B_3, and so θ_2 is not a simple extension of η.

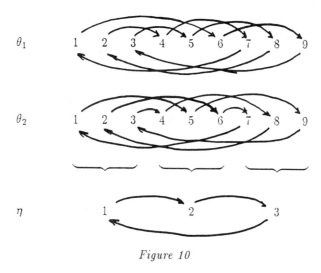

Figure 10

6.2. PROPOSITION. *Suppose $\theta \in \mathfrak{P}$ is an extension of $\eta \in \mathfrak{C}$ by $\alpha \in \mathfrak{P}$, and $\alpha \Rightarrow \beta \in \mathfrak{P}$. Then*

(i) *there exists $\widetilde{\theta} \in \mathfrak{P}$ such that $\theta \Rightarrow \widetilde{\theta}$ and $\widetilde{\theta}$ is an extension of η by β;*

(ii) *if θ is a simple extension, then $\widetilde{\theta}$ in (i) can be chosen to be a simple extension.*

PROOF:

Proof of (i): Let k (resp. m) denote the degree of η (resp. α). Suppose $\mathcal{B} = \{B_j\}_{j=1}^k$ is a block structure for θ over η such that $\theta^k|B_{j_1} = \{i + i_1 | i = 1, \ldots, m\}$ represents α. Let F be θ-adjusted on $\mathcal{P} = \{1, \ldots, n\}$, and let G be α-adjusted on $\mathcal{Q} = \{1, \ldots, m\}$. Since $\alpha \Rightarrow \beta$, there exists a representative $\mathcal{R}$ of β in G, with $\mathcal{R} \cap Flat(G, \mathcal{Q}) = \emptyset$. Note that F^k exhibits α on $B_{j_1} \subset \mathcal{P}$, so by *1.14* the ordered conjugacy $h : \mathcal{Q} \to B_{j_1}$ (between G and F^k, defined by $h(i) = i + i_1, i = 1, \ldots, m$) extends to an ordered conjugacy on $\mathcal{Q} \cap \mathcal{R}$. Since $\mathcal{R}$ is contained in the convex hull of $\mathcal{Q}$, $h[\mathcal{R}]$ is contained in the convex hull of B_{j_1}. But this means $h[\mathcal{R}] \subset \mathcal{R}(B_j, \mathcal{P})$, so that $h[\mathcal{R}]$ is a subset of the representative set $\mathcal{R}(\mathcal{B}, \mathcal{P})$ of $\mathcal{B}$ with respect to $\mathcal{P}$. It follows from *3.3* that the union $\widetilde{R}$ of F-orbits of points in $h[\mathcal{R}]$ represents an extension $\widetilde{\theta}$ of a subpattern $\widetilde{\eta}$ of η; but since η is a cycle, we must have $\widetilde{\eta} = \eta$. Since $F^k|h(\mathcal{R})$ represents β, we have that $\widetilde{\theta}$ is an extension of η by β.

Proof of (ii): Note that in the preceding argument, if θ is a simple extension of η, then for $j \neq j_0$, F is monotone on $\mathcal{R}(B_j, \mathcal{P})$. It follows that F is monotone on $\widetilde{\mathcal{R}} \cap \mathcal{R}(B_j, \mathcal{P})$ for $j \neq j_0$, hence $\widetilde{\theta}$ is a simple extension of η by α. ∎

A central result in the characterization of forcing-minimal cycles is the observation that if $\theta \in \mathfrak{C}$ is an extension of $\eta \in \mathfrak{C}$ by a " Štefan" cycle α (see [**ALS; BlC; C; H**] for definition), then θ forces a simple extension of η by α (or by the conjugate of α by a flip). Such a statement is false if the Štefan cycle α is replaced by an arbitrary cycle. We give two examples, in which the statement fails for different reasons.

6.3. EXAMPLE. *The 8-cycle*

$$\theta = (1\ 7\ 3\ 6\ 4\ 8\ 2\ 5)$$

(figure 11) is a 4-extension of $\pi = (1\ 2)$ *by* $\alpha = \alpha_1 = (1\ 3\ 4\ 2)$, *but* θ *forces no simple extension of* π *by* α.

(Note that α is conjugate to itself via a flip.)

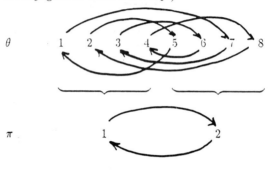

Figure 11

PROOF: Let F be θ-adjusted on $\mathcal{P} = \{1, \ldots, 8\}$. Then F has turning points at 2, 4, 5, and 6. Suppose $\mathcal{Q} = \{x_1, \ldots, x_8\}$ is a representative in F of some $\widetilde{\theta} \in \mathfrak{C}_8$ with a block structure $\mathcal{B}$ over $\pi = (1\ 2)$. The $\mathcal{B}$-intervals for $\widetilde{\theta}$ are $B_1 = [x_1, x_4]$ and $B_2 = [x_5, x_8]$; we must have $F[B_1] = B_2$ and $F[B_2] = B_1$. But this requires $B_1 \subset [1, 4]$ and $B_2 \subset [5, 8]$. The interior of each of these intervals contains only one turning point of F, so the restriction of $\widetilde{\theta}$ to each block is monotone or has a single turning point. However, α has two turning points, so if $\widetilde{\theta}^2|B_i = \alpha$ for $i = 1$ or $i = 2$, $\widetilde{\theta}$ cannot be monotone on either block; that is, it cannot be a simple extension of π by α. Thus F exhibits (and hence θ forces) no simple extension of α. ∎

The argument in the preceding example relies on the fact that α is not unimodal. Since every Štefan cycle is unimodal, this might tempt one to guess that an extension θ of π by any unimodal cycle α forces a simple extension of π by α. The next example shows this to be false, as well.

6.4. EXAMPLE. *There exists* $\theta \in \mathfrak{C}_{18}$ *which is a non-simple extension of* $\pi = (1\ 2\ 3)$ *by the unimodal cycle* $\alpha = (1\ 3\ 5\ 2\ 4\ 6)$ *and which forces no simple extension of* π *by* α.

PROOF: We construct θ (figure 12) by starting with the block structure $\mathcal{B} = \{B_j\}_{j=1}^3$, $B_j = \{6(j-1) + i \mid i = 1, \ldots, 6\}$; define $\theta : B_2 \to B_3$ so that $\theta(6 + i) = 12 + \alpha(i)$ for $i = 1, \ldots, 6$, and define $\theta : B_1 \to B_2$ (resp. $B_3 \to B_1$) by $\theta(i) = 6 + i$ (resp. $\theta(12 + i) = i$) unless $i \in \{1, 2\}\}$ (resp. $i \in \{3, 4\}\}$) in which case they are interchanged. This is sketched in the figure below, where the blocks are represented as columns and a copy of B_1 appears at either end.

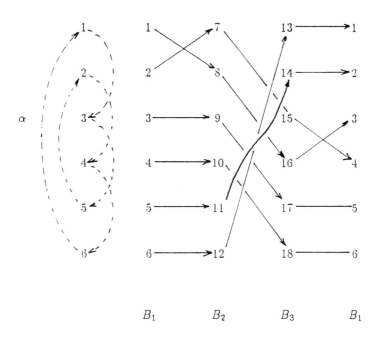

B_1 B_2 B_3 B_1

Figure 12

Since the α-image of $\{1, 2\}$ is $\{3, 4\}$, it is easily seen that $\alpha_1 = \alpha$. The reader can check that θ is the cycle

$$\theta = (1\ 8\ 16,\ 3\ 9\ 17,\ 5\ 11\ 14,\ 2\ 7\ 15,\ 4\ 10\ 18,\ 6\ 12\ 13).$$

Note that α is a doubling of $\beta = (1\ 2\ 3)$, so that θ is a doubling of $\eta \in \mathfrak{C}_9$, and by *6.1*, η is an extension of π by β. The block structure of θ over η is given by $\mathcal{D} = \{D_j\}_{j=1}^{9}$, where $D_j = \{2j - 1, 2j\}$.

We claim that θ forces no extension of π by α other than itself. To this end, we first show that η is the only 3-extension of π forced by θ. By *3.7(i)*, any cycle $\xi \neq \eta$ forced by θ either extends or is forced by η; if ξ is a 3-extension of π, then it cannot be an extension of η, so $\eta \Rightarrow \xi$. But it is easy to see that η is a simple extension of π by α, so as a corollary of *6.1(i)*, ξ must be an extension of π by some 3-cycle forced by β. Since β forces no 3-cycle other than itself, it follows that $\xi = \eta$, and we have that θ forces no 3-extension of π other than η.

Now, suppose $\widetilde{\theta}$ is an extension of π by α, with $\theta \Rightarrow \widetilde{\theta}$, and $\theta \neq \widetilde{\theta}$. Let F be the canonical θ-adjusted map, with $\mathcal{P} = \{1, \ldots, 18\}$. Let $\mathcal{Q} = \{x_1, \ldots, x_{18}\}$ be a representative of $\widetilde{\theta}$ in F. By *3.6(i)* $\mathcal{Q}$ is either contained in or disjoint from $\mathcal{R}(\mathcal{D}, \mathcal{P})$. Now since $\alpha \Rightarrow \beta$, *6.2(i)* tells us that $\widetilde{\theta}$ must force an extension of π by β; by the preceding argument, this extension must be η. Hence $\widetilde{\theta}$ cannot be forced by η (since $\widetilde{\theta} \Rightarrow \eta$ and forcing is antisymmetric on cycles). It follows from *3.6(ii)* that we must have $\mathcal{Q} \subset \mathcal{R}(\mathcal{D}, \mathcal{P})$. But the intervals $\mathcal{R}(D_j, \mathcal{P})$ are themselves $\mathcal{P}$-intervals, so this determines the itinerary of an orbit in $\mathcal{Q}$ to be the same as the itinerary of an orbit in $\mathcal{P}$. It follows from *1.5* that $\mathcal{Q}$ represents the same cycle as $\mathcal{P}$, that is, $\theta = \widetilde{\theta}$.

Hence, since θ is not a simple extension of π, the assertion is established. ∎

While example *6.4* shows that even an extension by a unimodal cycle α need not force a simple extension, the following result shows that when α is replaced by a 2-horseshoe, we do force a simple extension. A similar result plays a role in Coppel's study of forcing-minimal cycles ([C; B1C]).

6.5. PROPOSITION. *For* $\sigma = \pm 1$, *if* $\theta \in \mathfrak{P}$ *is an extension of* $\pi \in \mathfrak{C}$ *by* $\alpha = \mathfrak{h}(2, \sigma)$, *then* θ *forces a simple extension of* π *by* α.

PROOF:

Let $\mathcal{B} = \{B_j\}_{j=1}^k$ be a block structure for θ over π, with $\alpha_j = \mathfrak{h}(2, \sigma)$ for some j (which we can take to be $j = 1$, since we will not use the order between the blocks B_j). Let F be the canonical θ-adjusted map.

By *5.2*, the interval $\mathcal{R}(B_1, \mathcal{P})$ has subintervals $I_1 \leq I_2$ with disjoint interiors such that I_i F-covers $I_1 \cup I_2$ (I_1 with index σ) for $i = 1, 2$. We can assume that I_i minimally F^k-covers $I_1 \cup I_2$ and that no interval between I_1 and I_2 F^k-covers $I_1 \cup I_2$. For $j = 0, \ldots, k$ define

$$
\begin{aligned}
&I_i(j) = F^j(I_i), \quad i = 1, 2 \\
&I(j) = \text{convex hull } [I_1(j) \cup I_2(j)] \\
&J(j) = \text{clos } [I(j) \setminus \{I_1(j) \cup I_2(1)\}] \\
&A(j) = \partial I_1(j) \cup \partial I_2(j) \\
&B(j) = A(j) \setminus \partial I(j)
\end{aligned}
$$

Note that

(i) for $i = 1, 2$ and $j = 0, \ldots, k - 1$, $I_i(j)$ minimally F^{k-j}-covers $I(k)$;

(ii) $J(j)$ does not F-cover $I(j + 1)$;

(iii) $I(k) = I_1(k) = I_2(k)$ and $J(k) = \emptyset$;

(iv) $F[A(j)] = A(j + 1)$.

Now, if $F[B(j)] \subset B(j + 1)$, then F is monotone on $A(j)$; this is obvious if $B(j)$ is a single point (i.e., $J(j) = \emptyset$), and otherwise follows from *(i)*, since otherwise $\text{int}[I_i(j)]$ contains an F^{k-j}-preimage of $A(k)$. On the other hand, if a point of $B(j)$ maps to $\partial I(j + 1)$, then *(ii)* forces $B(j)$ to map to a single point, and this in turn forces $I_1(j + 1) = I_2(j + 1) = I(j + 1)$ and $J(j + 1) = \emptyset$; note that once this occurs for a given j, it remains true for all subsequent j, and F is monotone on $A(j + 1)$ for all these j.

We have shown that if j_0 is the highest value of j for which $I_1(j) \neq I_2(j)$, then for $j < j_0$, $F[B_j] \subset B(j + 1)$ and so F maps $A(j)$ monotonically to $A(j + 1)$, while for $j > j_0$, $A(j) = \partial I(j)$ is mapped monotonically to $A(j + 1)$. Now, let $C \subset A(0)$ consist of both endpoints of $I(0)$ and one point of $B(0)$. Then for $j \neq j_0$, F maps $F^j[C]$ monotonically onto $F^{j+1}[C]$, and F^k exhibits $\alpha = \mathfrak{h}(2, \sigma)$ on C. Thus the set $\bigcup_{j=0}^{k} F^j(C)$ represents a simple extension of π by α in F, proving the result. ∎

An immediate corollary of *5.4* and *6.5*, via *6.2 (iii)*, is the following

6.6. COROLLARY. *Suppose* $\theta \in \mathfrak{P}$ *is an extension of* $\pi \in \mathfrak{C}$ *by* α, *where* $\alpha = \mathfrak{h}(k, \sigma)$, *and* $\beta \in \mathfrak{C}$ *is a* $(2, \sigma')$-fold cycle. *If* $k \geq 3$ *or* $k = 2$ *and* $\sigma = \sigma'$, *then* θ *forces a simple extension of* π *by* β.

7. Combinatorial Shadowing

Our results so far have used all the combinatorial data obtainable from a finite invariant set, encoded in a combinatorial pattern, to determine the cycles, permutations and patterns it forces. In this section, we formulate a result which allows us to conclude a forcing relation $\theta \Rightarrow \xi$ from only partial knowledge of θ. The idea is that, starting from a known strong forcing relation $\eta \Rightarrow\!\!\!\succ \xi$, we look for a sufficiently large piece of θ which "mimics" η; if we find it, we can conclude that $\theta \Rightarrow \xi$. The piece need not be invariant–indeed, it may be a finite piece of an infinite orbit or invariant set–but it must include sufficiently long pieces of orbits.

Invariant sets are modelled combinatorially by combinatorial patterns; to model a part of an invariant set, itself not necessarily invariant, we formulate the following.

7.1. DEFINITION. *A* **semi-pattern on** n **elements** *is a map*

$$\theta : \mathcal{D}(\theta) \to \mathcal{P}_n = \{1, \ldots, n\},$$

where $\mathcal{D}(\theta) \subset \mathcal{P}_n$, *such that every point of* $\mathcal{P}_n$ *belongs either to the domain or the range of* θ:

$$\mathcal{D}(\theta) \cup \theta[\mathcal{D}(\theta)] = \mathcal{P}_n.$$

We denote the collection of all semi-patterns (resp. semi-patterns on n elements) by Λ (resp. Λ_n). Given a semi-pattern, θ, we can define the iterates θ^i of the underlying map θ on the domains $\mathcal{D}_i(\theta)$ inductively by

$$\mathcal{D}_1(\theta) = \mathcal{D}(\theta),$$

$$\mathcal{D}_{i+1}(\theta) = \{x \in \mathcal{D}_i(\theta) \mid \theta(x) \in \mathcal{D}_i(\theta)\}$$

and $\theta^{i+1}(x) = \theta^i(\theta(x))$, as long as $\mathcal{D}_i(\theta) \neq \emptyset$.

Using these iterates, we single out a special class of semi-patterns, modelling pieces of one orbit.

7.2. DEFINITION. *A semi-pattern* $\theta \in \Lambda_n$ *is an* **orbit segment** *of* **length** n *if there exists* $x \in \mathcal{D}_n(\theta)$ *such that for each* $y \in \mathcal{P}_n = \{1, \ldots, n\}$ *there is* $i = 1, \ldots, n$ *with* $y = \theta^i(x)$.

Note that the domain $\mathcal{D}(\theta)$ of an orbit segment θ excludes at most one point of $\mathcal{P}_n$ (namely $\theta^n(x)$). We can think of an orbit segment as a sequence $z_i \in \{1, \ldots, n\}$, $i = 0, \ldots, n-1$ such that $z_{i+1} \in \theta(z_i)$. (Here the index refers to temporal, rather than spatial, order.)

The following naturally extends block structures to semi-patterns.

7.3. DEFINITION. *Suppose* $\theta \in \Lambda$ *and* $\eta \in \Lambda_k$. *A* **block structure for** θ **over** η *is a partition* $\mathcal{B} = \{B_j\}_{j=1}^k$ *of* $\mathcal{P}_n = \{1, \ldots, n\}$ *into blocks, where the blocks are numbered left-to-right:*

(i) $x \in B_i, y \in B_j$ *and* $i < j$ *implies* $x < y$

such that for each $i = 1, \ldots, k$,

(ii) $\theta[B_i \cap \mathcal{D}(\theta)] \subset B_{\eta(i)}$.

Note that if a block B_i is disjoint from the domain $\mathcal{D}(\theta)$, then $i \notin \mathcal{D}(\eta)$ and condition *(ii)* is vacuous. We shall be interested only in situations where this does not occur.

7.4. DEFINITION. *Let* $m \in \mathbb{N}$. *We say* $\theta \in \Lambda$ *m-shadows* $\eta \in \Lambda_k$ *if* θ *has some block structure* $\mathcal{B} = \{B_j\}_{j=1}^{k}$ *over* η *such that*

$$B_i \cap \mathcal{D}_m(\theta) \neq \emptyset \text{ for } i = 1, \ldots, k.$$

We say simply that θ **shadows** η if θ 1-shadows η.

7.5. REMARK.

(i) *If a pattern* $\theta \in \mathfrak{P}_n$ *has a block structure over* $\eta \in \Lambda_k$, *then* $\eta \in \mathfrak{P}_k$ *and definition 3.1 applies:* θ *is an extension of* η, *and (equivalently)* θ *m-shadows* η *for all* $m \in \mathbb{N}$.

(ii) *More generally, if* $\theta \in \Lambda_n$ *shadows* $\eta \in \Lambda_k$, *then by* 7.4, $\mathcal{D}(\eta) = \mathcal{P}_k$, *so* $\eta \in \mathfrak{P}_k$.

(iii) *If* $\theta \in \Lambda_n$ *is an orbit segment of length* n, *then a block structure for* θ *over* $\eta \in \Lambda_k$, $\theta \neq \eta$, *must imply that* η *is a k-cycle,* $n > k$, *and* θ $(n-k)$-*shadows* η.

(iv) *Suppose* $\theta \in \Lambda$ *m-shadows a pattern* $\eta \in \mathfrak{P}_n$, *and let* $\mathcal{B} = \{B_j\}_{j=1}^{n}$ *be the appropriate block structure for* θ *over* η. *The m-shadowing hypothesis means each block* $B_j, j = 1, \ldots, n$ *contains at least one point* $x_j^0 \in B_j$ *such that* x_j^i *is defined for* $i = 1, \ldots, m$ *by* $x_j^{i+1} = \theta(x_j^i)$ *and* $x_j^i \in B_{\eta^i(j)}$. *The (not necessarily disjoint) union of the orbit segments* $\{x_j^i \mid i = 0, \ldots, m\}$ *is a sub-semi-pattern of* θ *which also m-shadows* η.

The basic idea of our *Combinatorial Shadowing Theorem* is this: suppose $\xi \in \mathfrak{S}_k(\eta)$, where $\eta \in \mathfrak{P}_n$. Then there is an estimate $N = N(n, k)$ such that any semi-pattern θ which m-shadows η with $m \geq N$ must force ξ. However, there are some difficulties with this statement if ξ includes some subcycles of η; we need to assume that η strongly forces ξ. We reformulate the definition of this notion from the introduction.

7.6. DEFINITION. *Given patterns* $\theta \in \mathfrak{P}_n$ *and* $\eta \in \mathfrak{P}_k$, *we say* θ **strongly forces** η *(denoted* $\theta \Rightarrow\!\!\!\succ \eta$) *if there exists a pattern* $\theta \vee \eta \in \mathfrak{P}_{n+k}$ *and ordered conjugacies* $h_\theta : \mathcal{P}_n \to \mathcal{P}_{n+k}$, $h_\eta : \mathcal{P}_k \to \mathcal{P}_{n+k}$ *such that*

(i) $\theta \Rightarrow \theta \vee \eta$;
(ii) $(\theta \vee \eta) \circ h_\theta = h_\theta \circ \theta$;
(iii) $(\theta \vee \eta) \circ h_\eta = h_\eta \circ \eta$;
(iv) $h_\theta[\mathcal{P}_n] \cap h_\eta[\mathcal{P}_k] = \emptyset$.

Notice that, unlike ordinary forcing, strong forcing is not transitive. For example, let $\theta = (1\ 2\ 4)(3)$, $\eta = (1\ 2)$, and $\pi = (1)$ (see figure 13).

Then $\theta \Rightarrow\!\!\!\succ \eta$, with $\theta \vee \eta = (1\ 2\ 6)(3\ 5)(4)$, and $\eta \Rightarrow\!\!\!\succ \pi$, with $\eta \vee \pi = (1\ 3)(2)$, but $\theta \not\Rightarrow\!\!\!\succ \pi$ (since a θ-adjusted map has only one fixedpoint). We shall use a star to denote the set of items strongly forced by a given one (thus, $\mathfrak{P}^*(\theta)$ denotes the patterns strongly forced by θ, etc.)

An easy corollary of 1.21 is the following characterization of strong forcing:

7.7. LEMMA. *Given* θ, $\eta \in \mathfrak{P}$ *and* F θ-adjusted on $\mathcal{P}$, *then* $\theta \Rightarrow\!\!\!\succ \eta$ *if and only if there exists a representative* $\mathcal{Q}$ *of* η *in* F *with*

$$\mathcal{Q} \cap [\mathcal{P} \cup Flat(F, \mathcal{P})] = \emptyset.$$

PROOF: If there exists such a representative, then $\mathcal{R} = \mathcal{P} \cup \mathcal{Q}$ represents $\theta \vee \eta$ and is disjoint from $Flat(F, \mathcal{P})$, so $\theta \Rightarrow \theta \vee \eta$, as required.

Conversely, suppose $\theta \Rightarrow \theta \vee \eta$ as in 7.6 and let $\mathcal{R}$ represent $\theta \vee \eta$ with $\mathcal{R} \cap Flat(F, \mathcal{P}) = \emptyset$. Write $\mathcal{R} = \mathcal{P}' \cup \mathcal{Q}$, where $\mathcal{P}'$ and $\mathcal{Q}$ correspond to the images

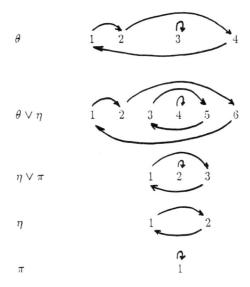

Figure 13

of h_θ and h_η, respectively. Then $\mathcal{P}' \cap \mathcal{Q} = \emptyset$, and $\mathcal{P}'$ is a representative of θ in F with $\mathcal{P}' \cap Flat(F, \mathcal{P}) = \emptyset$; by *1.19*, we must have $\mathcal{P}' = \mathcal{P}$. Thus, $\mathcal{Q}$ is disjoint from both $Flat(F, \mathcal{P})$ (since $\mathcal{Q} \subset \mathcal{R}$) and from $\mathcal{P}' = \mathcal{P}$. ∎

Two conclusions follow easily from *7.7*.

7.8. LEMMA.

  (i) For any $\theta \in \mathfrak{P}$, $\theta \not\Rrightarrow \theta$.
  (ii) For θ, $\eta \in \mathfrak{C}$, $\theta \Rrightarrow \eta$ iff $\theta \Rightarrow \eta$ and $\theta \neq \eta$.

PROOF: *(i)* is immediate from *7.7* and *1.19*.

Proof of (ii): If $\theta \Rrightarrow \eta$, then $\theta \Rightarrow \theta \vee \eta \Rightarrow \eta$, so $\theta \Rightarrow \eta$, and $\theta \neq \eta$ by *(i)*. Conversely, suppose $\theta \Rightarrow \eta$, θ, $\eta \in \mathfrak{C}$, and $\theta \neq \eta$. Let F be θ-adjusted on $\mathcal{P}$ and pick $\mathcal{Q}$ a representative of η in F. Since $\theta \in \mathfrak{C}$, $Flat(F, \mathcal{P}) = \emptyset$. Furthermore, since $\mathcal{P}$ and $\mathcal{Q}$ each consist of a single orbit, either $\mathcal{P} = \mathcal{Q}$ or $\mathcal{P} \cap \mathcal{Q} = \emptyset$. But in the first case, $\theta = \eta$ contrary to hypothesis. Thus, $\mathcal{Q} \cap [\mathcal{P} \cup Flat(F, \mathcal{P})] = \emptyset$, and $\theta \Rrightarrow \eta$ by *7.7*. ∎

We note *7.8(ii)* is no longer true when θ is a permutation: a permutation may strongly force one of its subcycles. For example(figure 14), let $\theta \in \mathfrak{S}_5$ be $\theta = (1\ 5)(2\ 3\ 4)$, and $\eta = (1\ 2)$. Then $\eta \subseteq \theta$, but the canonical θ-adjusted map also has a periodic orbit of period 2 in the interval $(2, 4)$. Thus, $\theta \Rightarrow \theta \vee \eta = (1\ 7)(2\ 4\ 6)(3\ 5)$, hence $\theta \Rrightarrow \eta$.

With this background, we can begin proving a sequence of combinatorial shadowing results. The first result says that an orbit segment θ which $3n$-shadows a doubling η of some n-cycle π forces π. We formulate this statement carefully for future use.

7.9. PROPOSITION. *Suppose $f \in \mathcal{E}$ and $\theta \in \Lambda_{3n}$ is an orbit segment of length $3n$ exhibited by f on the set $\mathcal{P} = \{x, f(x), f^2(x), \ldots, f^{n-1}(x)\}$. Suppose $\eta \in \mathfrak{C}_{2n}$ is a doubling of $\pi \in \mathfrak{C}_n$, θ shadows η and $\mathcal{B} = \{B_j\}_{j=1}^{2n}$ is a block structure for θ over η.*

Let D_i, $i = 1, \ldots, n$ be the convex hull of the points of $\mathcal{P}$ in $B_{2i-1} \cup B_{2i}$. Then there exists a representative of π intersecting each D_i in one point. In particular,

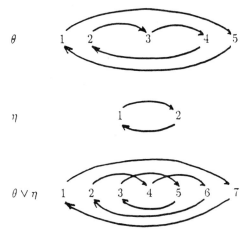

θ

η

$\theta \vee \eta$

Figure 14

θ *forces* η.

PROOF: In this proof, we regard θ as the sequence $\{x_i\}_{i=1}^{3n}$ defined by $x_i = f^{i-1}(x)$ for $i = 1, \ldots, 3n$. Our proof will be by contradiction; suppose there is no $z \in D_1$ such that $f^j(z) \in D_{\pi^j(1)}$ for $j = 1, \ldots, n$ and $f^n(z) = z$.

CLAIM 1: *There is no proper loop consisting of subintervals* $J_k \subset D_{\pi^k(1)}$, *for* $k = 0, \ldots, n$, *where* $J_n = J_0$.

This is simply a consequence of the fact that any such loop would be the itinerary of the orbit $\mathcal{O}(z)$ which we have assumed is nonexistent.

CLAIM 2: *For* $0 \leq i \leq n-1$, $x_{i+2n} \in \langle x_i, x_{i+n} \rangle$.

The other possibilities are

    (a) $x_{i+n} \in \langle x_i, x_{i+2n} \rangle$
    (b) $x_i \in \text{clos} \langle x_{i+n}, x_{i+2n} \rangle$
    (c) $x_{i+n} = x_{i+2n}$.

Let $I_j = \text{clos} \langle x_{i+j}, x_{i+j+n} \rangle$ for $j = 0, \ldots, n-1$. Clearly, I_j f-covers I_{j+1} for $j < n$, so $\mathcal{I} = \{I_j\}_{j=0}^{n-1}$ is a proper itinerary. Since x_i and x_{i+2n} belong to the same B_k, and x_{i+n} does not, case *(a)* is ruled out. In case *(b)*, I_{n-1} f-covers I_0, so I is a loop, contrary to claim *1*. In case *(c)*, $\mathcal{O}(x)$ is a representative of π in the D's, contrary to assumption.

We now use f to extend the sequence x_i to all positive integer indices, by $x_{i+1} = f(x_i)$. Given $i \in \mathbb{Z}^+$, the Euclidean algorithm gives unique integers $\alpha(i)$, $\beta(i)$ such that

$$i = \alpha(i) + n\beta(i), \quad 0 \leq \alpha(i) \leq n-1.$$

Set

$$A_i = \{x_{\alpha(i)+nk} \mid 2 \leq k \leq \beta(i)\}$$
$$E_i = \{x_{\alpha(i)+nk} \mid 0 \leq k \leq \beta(i)\}$$

CLAIM 3: *For every* $i > 2n$,

    (a) $A_i \subset \langle x_{\alpha(i)}, x_{\alpha(i)+n} \rangle$
    (b) *given* u *in the convex hull of* A_i, *there exist* $y, z \in E_i$ *with* $y < u \leq f^n(y)$
        and $z > u \geq f^n(z)$.

We prove claim *3* by induction on i, starting with the case $2n \leq i \leq 3n$. In this case, A_i consists of a single point, and the claim follows immediately from claim *2*.

Now, suppose $j > 3n$ and assume *(a)*, *(b)* hold for all i with $2n < i < j$. We prove *(a)* for $i = j$ by contradiction. Suppose

$$x_j \notin \langle x_{\alpha(j)}, x_{\alpha(j)+n} \rangle.$$

Using *(b)* for $i = j - n$, we find $v \in E_{j-n}$ such that either

$$x_j \leq v \leq x_{j-n} \leq f^n(v)$$

or

$$x_j \geq v > x_{j-n} \geq f^n(v).$$

For $m = 0, \ldots, n - 1$, the intervals

$$I_m = \text{clos}\langle f^m(v), x_{j-n+m} \rangle$$

form a proper loop (I_m f-covers I_{m+1} by the condition on v). By *(a)* for $i = j - n, \ldots, j - 1$, each I_m is contained in some D_k, and we have a contradiction. Thus, *(a)* is satisfied for $i = j$.

Now, to show *(b)* for $i = j$, note that if x_j belongs to the convex hull of A_{j-n}, then A_j and A_{j-n} have the same convex hull, so *(b)* follows for $i = j$ from the case $i = j - n$. On the other hand, if x_j is not in the convex hull of A_{j-n}, we have two possible configurations (by claim *2*):

(i) $x_j \in \langle x_{\alpha(j)+n}, x_{\alpha(j)+2n} \rangle$
(ii) $x_j \in \langle x_{\alpha(j)}, x_{\alpha(j)+2n} \rangle$.

In case *(i)*, take $\{y, z\} = \{x_{\alpha(j)}, x_{j-n}\}$. this shows *(b)* for $i = j$, and completes the proof of claim *3*.

Now to complete the proof, for $0 \leq i \leq n - 1$, set

$$I_i = [\liminf_{k \to \infty} x_{i+kn}, \limsup_{k \to \infty} x_{i+kn}].$$

Certainly, I_i f-covers I_{i+1} for $i < n - 1$, and I_{n-1} covers I_0. Furthermore, by *3(a)* each I_i is contained in some D_j, contradicting claim *1* and proving the proposition. ∎

The main result of this section is the following.

7.10. THEOREM. (**Combinatorial Shadowing Theorem**) *Suppose η is a pattern which strongly forces the permutation $\pi \in \mathfrak{S}$. Let*

$L = \max[\text{length of subcycles of } \pi]$
$M = 2 \cdot \max[\text{least common multiples of lengths of pairs of subcycles of } \pi]$.
Set $N = M + 5L - 1$.

Then any semi-pattern θ which N-shadows η must force π.

An immediate corollary to this theorem, in view of *7.5* and *7.8*, is the following.

7.11. COROLLARY. *Suppose $\eta \in \mathfrak{C}_k$, $\pi \in \mathfrak{C}_n$, $\pi \neq \eta$ and η forces π. Then*

(i) *any orbit segment of length $(k + 7n - 2)$ with a block structure over η forces π*
(ii) *any pattern which shadows η forces π.*

Proof of (7.10): Our proof follows the proofs of *1.10* and *1.14*.

Given $\theta \in \Lambda_s$, N-shadowing $\eta \in \mathfrak{P}_t$, and $\eta \Rightarrow\!\!\!\!> \pi \in \mathfrak{S}_n$, let $\mathcal{B} = \{B_j\}_{j=1}^t$ be a block structure for θ over η satisfying the definition 7.4 of m-shadowing with $m = N$.

Let F be η-adjusted on $\mathcal{Q}$; let

$$\mathcal{Q} = \mathcal{Q}_0 = \{y_1^0 < y_2^0 < \cdots < y_{t(0)=t}^0\}$$

and as in *1.19* recursively define

$$\mathcal{Q}_\ell = \{y_1^\ell < y_2^\ell < \cdots < y_{t(\ell)}^\ell\}$$

by

$$\mathcal{Q}_{\ell+1} = \{x \notin Flat(F, \mathcal{Q}_\ell) \mid F(x) \in \mathcal{Q}_\ell\}.$$

Also, suppose $f \in \mathcal{E}(I)$ exhibits θ on

$$\mathcal{P}_0 = \{x_1 < x_2 < \cdots < x_s\};$$

let

$$A_0 = \{x_j \in \mathcal{P}_0 \mid j \in D(\theta)\}$$

and for $i = 1, \ldots, k$,

$$B_i^0 = \{x_j \in \mathcal{P}_0 \mid j \in B_i\}.$$

To show that f exhibits π, we first establish the following claim.

CLAIM : *For $k = 0, 1, \ldots$ there exist sets A_k and $\mathcal{P}_k$ and disjoint blocks B_i^k, $i = 1, \ldots, t(k)$, such that $\mathcal{P}_0$, A_0 and B_i^0 ($i = 1, \ldots, t(0) = t$) agree with the sets above and the following conditions hold:*

(i) *if $x \in B_i^k$, $y \in B_j^k$ and $i < j$ then $x < y$;*

(ii) *if $x \in B_i^k \cap A_k$ and $f(x) \in B_j^k$, then*

$$F(y_i^k) = y_j^k;$$

(iii) *for every $i \in \{1, \ldots, t(k)\}$ there exists $x \in B_i^k$ such that*

$$x, f(x), \ldots, f^{N-k-1}(x) \in A_k;$$

(iv) *$A_k \subset \mathcal{P}_k = B_1^k \cup \cdots \cup B_{t(k)}^k$.*

Proof of claim: We will prove the claim by induction on k. The conditions clearly hold for $k = 0$.

Now let $\ell < N$ and assume $\mathcal{P}_\ell = \{B_1^\ell \cup \cdots \cup B_{t(\ell)}^\ell\}$ and A_ℓ are defined and satisfy *(i)-(iv)* with $k = \ell$. We need to define $\mathcal{P}_{\ell+1} = \{B_1^{\ell+1} \cup \cdots \cup B_{t(\ell+1)}^{\ell+1}\}$ and $A_{\ell+1}$ so *(i)-(iv)* hold with $k = \ell + 1$.

To construct $\mathcal{P}_{\ell+1} = B_1^{\ell+1} \cup \cdots \cup B_{t(\ell+1)}^{\ell+1}$, we first examine the points $y_i^{\ell+1}$ in the set $\mathcal{Q}_{\ell+1}$ (already defined). First, for each $y_i^{\ell+1} \in \mathcal{Q}_{\ell+1}$ which belongs to $\mathcal{Q}_\ell$, find the unique $y_j^\ell \in \mathcal{Q}_\ell$ which it equals and set

$$B_i^{\ell+1} = B_j^\ell \cap A_\ell.$$

Now, suppose the $\mathcal{Q}_\ell$-interval $I_j = (y_j^\ell, y_{j+1}^\ell)$ contains other points of $\mathcal{Q}_{\ell+1}$, say

$$\mathcal{Q}_{\ell+1} \cap I_j = \{y_r^{\ell+1} < \cdots < y_m^{\ell+1}\}.$$

(Note that this means $y_{r-1}^{\ell+1} = y_j^\ell$ and $y_{m+1}^{\ell+1} = y_{j+1}^\ell$.) Then F is strictly monotone on clos I_j, so that

$$Q_\ell \cap \langle F(y_j^\ell), F(y_{j+1}^\ell) \rangle = F[Q_{\ell+1} \cap I_j]$$
$$= \{F(y_r^{\ell+1}) <_\sigma \dots <_\sigma F(y_m^{\ell+1})\},$$

where $\sigma \in \{\pm 1\}$ is determined by

$$F(y_{r-1}^{\ell+1}) = F(y_j^\ell) <_\sigma F(y_{j+1}^{\ell+1}).$$

But the points of $Q_\ell \cap F[I_j]$ uniquely identify blocks for $\mathcal{P}_\ell$; let W be their union and J its convex hull. Also, since $y_{r-1}^{\ell+1}$ and $y_{m+1}^{\ell+1}$ belong to Q_ℓ, the blocks $B_{r-1}^{\ell+1}$ and $B_{m+1}^{\ell+1}$ are well defined, and we can set $I = (\max[B_{r-1}^{\ell+1}], \min[B_{m+1}^{\ell+1}])$, and apply lemma *1.1(i)* to find a finite set $V \subset I$ which maps σ-increasing onto W under f. This set divides into blocks $B_r^{\ell+1}, \dots, B_m^{\ell+1}$ in a natural way, and we adjoin it to the part of $\mathcal{P}_{\ell+1}$ already defined as $\mathcal{P}_{\ell+1} \cap \mathcal{P}_\ell$.

The above construction gives $\mathcal{P}_{\ell+1}$ and the blocks $B_i^{\ell+1}$. Now define $A_{\ell+1}$ by: $x \in A_{\ell+1}$ iff there exist i, j such that

$$x \in B_i^{\ell+1}, \ f(x) \in B_j^{\ell+1}$$

and $F(y_i^{\ell+1}) = y_j^{\ell+1}$.

With this definition, we clearly have *(i)*, *(ii)* and *(iv)* for $k = \ell + 1$; it remains to check *(iii)*. Pick $i \in \{1, \dots, t(\ell+1)\}$.

If $y_i^{\ell+1} = y_j^\ell$ for some j, we have by inductive hypothesis that some $x \in B_j^\ell$ satisfies

$$x, f(x), \dots, f^{N-(\ell+1)-1}(x) \in A_{\ell+1}.$$

Now, for $r = 0, \dots, N - \ell - 1$, the point $f^r(x)$ belongs to $A_\ell \cap B_{\varphi(r)}^\ell$ for some $\varphi(r)$, and hence by construction to some $B_{\psi(r)}^{\ell+1}$. Furthermore, this construction insures that $y_{\varphi(r)}^\ell = y_{\psi(r)}^{\ell+1}$ and the inductive hypothesis gives us

$$F(y_{\varphi(r)}^\ell) = y_{\varphi(r+1)}^\ell \quad \text{for } r = 0, \dots, N - \ell - 2.$$

But we can replace ℓ with $\ell + 1$ and φ with ψ in this last equation to show (by the definition of $A_{\ell+1}$) that

$$x, f(x), \dots, f^{N-\ell-2}(x) \in A_{\ell+1}.$$

Since $N - \ell - 2 = N - (\ell+1) - 1$, we have *(iii)* in this case.

If $y_i^{\ell+1} \notin Q_\ell$, we have p such that

$$F(y_i^{\ell+1}) = y_p^{\ell+1} \in Q_\ell.$$

The above arguments then give us $y \in B_q^{\ell+1} \cap A_{\ell+1}$ such that

$$y, f(y), \dots, f^{N-(\ell+1)-1}(y) \in A_{\ell+1}.$$

But by the earlier construction there is $x \in B_j^{\ell+1}$ with

$$f(x) = y$$

and by definition $x \in A_{\ell+1}$. Thus,

$$x, f(x), \ldots, f^{N-(\ell+1)}(x) \in A_{\ell+1}$$

and we have *(iii)* for $k = \ell + 1$ in this case.

This completes the induction, and proves the claim.

Now, we need to produce a representative of π in f. Since η strongly forces π and F is η-adjusted on $\mathcal{Q}$, by *7.7* we have a representative $\mathcal{R}$ of π in F with $\mathcal{R} \cap [\mathcal{Q} \cup Flat(F, \mathcal{Q})] = \emptyset$. We will (carefully) construct a periodic f-orbit corresponding to each (periodic) F-orbit in $\mathcal{R}$.

Pick $x \in \mathcal{R}$ and let K be its least period under F. Notice that $K \leq L$, and set

$$H = M + 2K.$$

Consider the $\mathcal{Q}_H$-interval (y_i^H, y_{i+1}^H) containing x.

If this interval is self-tandem, we can apply proposition *7.9*, since $N \geq 3K - 1 + H$ and parts *(ii)* and *(iii)* of the claim insure that for some $u \in B_i^H$ the orbit segment $f^j(u)$, $j = 0, \ldots, 3K - 1$ shadows the (self-tandem) cycle given by the F-orbit of y_i^H. We obtain a periodic f-orbit $\{z_j\}_{j=0}^{k-1}$ of least period K with

$$f(z_j) = z_{j+1}, \ f(z_{k-1}) = z_0$$

and z_j contained in the convex hull of $B_{\varphi(j)}^H \cup B_{\varphi(j+k)}^H$ (where $F^j(y_i^H) = y_{\varphi(j)}^H$). Note that here $B_{\varphi(j)}^H$ and $B_{\varphi(j+k)}^H$ are adjacent blocks and

$$y_{\varphi(j)}^H, y_{\varphi(j+k)}^H \in \mathcal{Q} \subset \mathcal{Q}_M.$$

If the interval (y_i^H, y_{i+1}^H) is not self-tandem, then lemma *1.11* with $\mathcal{P} = \mathcal{Q}_M$ applies, as follows. Since $N \geq H + 2K$, parts *(ii)* and *(iii)* of the claim give us points $u \in B_i^H$, $v \in B_{i+1}^H$ such that

$$f^j(u) \in B_{\varphi(j)}^H, \ f^j(v) \in B_{\psi(j)}^H$$

for $j = 0, \ldots, 2K$, where φ, ψ are determined by

$$F^j(y_i^H) = y_{\varphi(j)}^H$$

and

$$F^j(y_{i+1}^H) = y_{\psi(j)}^H,$$

respectively. Clearly for $j = 0, \ldots, 2K - 1$ the interval

$$\mathrm{clos}\langle f^j(u), f^j(v)\rangle$$

f-covers the interval

$$\mathrm{clos}\langle f^{j+1}(u), f^{j+1}(v)\rangle.$$

Lemma *1.11* insures that the blocks B_i^H and B_j^H lie strictly between the blocks $B_{\varphi(2K)}^H$ and $B_{\psi(2K)}^H$. Therefore

$$\mathrm{clos}\langle f^{2K}(u), f^{2K}(v)\rangle \supset \mathrm{clos}\langle u, v\rangle,$$

so that the itinerary $\{\mathrm{clos}\langle f^j(u), f^j(v)\rangle\}_{j=0}^{2K-1}$ is a proper loop. Thus we have a periodic f-orbit $\{w_j\}_{j=0}^{2K-1}$ with $f(w_j) = w_{j+1}$, $f(w_{2K-1}) = w_0$ such that w_j

belongs to the convex hull of $B_{\varphi(j)}^H \cup B_{\psi(j)}^H$. In fact, lemma *1.11* gives us more information. If the $\mathcal{Q}_M$-interval containing $F^j(x)$ is $(y_{\lambda(j)}^H, y_{\mu(j)}^H)$, then we know that $B_{\varphi(j)}^H \cup B_{\psi(j)}^H$ is contained in the convex hull of $B_{\lambda(j)}^H \cup B_{\mu(j)}^H$, and so for $j = 0, \ldots, 2K - 1$, w_j belongs to the convex hull of $B_{\lambda(j)}^H \cup B_{\mu(j)}^H$. Note that $B_{\lambda(j+k)}^H = B_{\lambda(j)}^H$ and $B_{\mu(j+k)}^H = B_{\mu(j)}^H$.

By proposition *1.13*, if $i \neq j \in \{0, \ldots, K-1\}$, then there are at least two points of $\mathcal{Q}_M$ between $F^i(x)$ and $F^j(x)$. Therefore if i, $j \in \{0, \ldots, 2K-1\}$ are distinct and do not differ by K, then the convex hulls of $B_{\lambda(i)}^H \cup B_{\lambda(i)}^H$ and of $B_{\lambda(j)}^H \cup B_{\lambda(j)}^H$ are disjoint. Hence $\{w_i\}_{i=0}^{2K-1}$ represents either the cycle exhibited by F on the orbit of x, or its doubling. In the first case, $w_k = w_0$ and we let $z_i = w_i$, $i = 0, \ldots, K-1$; in the second case f has a periodic orbit $\{z_i\}_{i=0}^{K-1}$ with $z_i \in \langle w_i, w_{i+K} \rangle$. In both cases this orbit is ordered-conjugate to that of x, and z_i belongs to the convex hull of $B_{\lambda(i)}^H \cup B_{\mu(i)}^H$.

Now to complete the proof of the theorem, we carry out this construction for each periodic F-orbit in $\mathcal{R}$ (picking one x in turn from each such orbit). The correspondence defined on each orbit $\mathcal{O}(x)$ by $f^j(x) \mapsto z_j$ gives a map $h : \mathcal{R} \to I$ such that $f \circ h = h \circ F$. Proposition *1.13* insures that no $\mathcal{Q}_M$-interval intersects $\mathcal{R}$ in more than one point; thus, if $x, y \in \mathcal{R}$ belong to distinct F-orbits in $\mathcal{R}$ then $h(x)$ and $h(y)$ belong to disjoint corresponding sets of the form "convex hull of $(B_{\lambda(i)}^H \cup (B_{\mu(i)}^H)$". Thus, h is an ordered conjugacy, and $h[\mathcal{R}]$ is a representative of π in f, as required. ∎

The combinatorial shadowing theorem tells us that a semi-pattern θ which m-shadows a pattern η is guaranteed, provided m is sufficiently large, to force $\mathfrak{S}_n^*(\eta)$. Note that the estimate on m depends only on n, not on the pattern η. As m gets large, we can guarantee that θ forces more and more of $\mathfrak{S}^*(\eta)$, so that "in the limit" we expect a map exhibiting semi-patterns which m-shadow η for all m to also exhibit η. The following example shows that this need not be true if η has self-tandem cycles.

7.12. EXAMPLE. *Let $f : [1, 5] \to [1, 5]$ be the piecewise-linear map defined by (see figure 15)*

$$f(1) = 5, \quad f(3) = 4, \quad f(4) = 2, f(5) = 3.$$

(This implies $f(2) = 4\frac{1}{2}$.) Then f exhibits orbit segments of arbitrary length which shadow the cycle $\eta = (1\ 4\ 2\ 3)$, but f does not exhibit η.

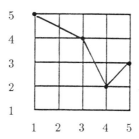

Figure 15

PROOF: Note that the interval $[1, 3]$ maps affinely onto $[4, 5]$, and $[4, 5]$ maps linearly onto $[2, 3] \subset [1, 3]$. Thus, any orbit intersecting $[1, 3] \cup [4, 5]$ is trapped in this union. Since $[3, 4]$ maps onto $[2, 4]$ with slope -2, the only periodic orbit intersecting $[3, 4]$ is the fixedpoint at $\frac{10}{3}$. On the other hand, the slope of f is $-\frac{1}{2}$ on $[1, 3]$ and 1 on $[4, 5]$, so this union is contained in the basin of attraction of the periodic attractor of period 2, $\{\frac{7}{3}, \frac{13}{3}\}$. Hence these are the only periodic orbits of f; in particular, f cannot exhibit η.

On the other hand, the f-orbit of 1 spirals in toward this attractor, and so any initial segment shadows η. ∎

Two aspects of 7.12 deserve comment. First, f cannot be θ-adjusted for any pattern θ, since such maps cannot have attracting non-critical periodic orbits. However, if one follows the f-orbit of 1 arbitrarily long until it ends up in $[4, 5]$, say $f^{2n+1}(1) = y$, then one can change f on the open interval $\langle f^{2n-3}(1), \frac{13}{3}\rangle$ (which contains y) by collapsing $f[\langle y, \frac{13}{3}\rangle]$ to the point $\frac{7}{3}$ and mapping the interval $\langle f^{2n-3}(1), y\rangle$ affinely onto $\langle f^{2n-2}(1), \frac{7}{3}\rangle$. This new map is θ-adjusted for the pattern θ represented by the 2-cycle $\{\frac{7}{3}, \frac{13}{3}\}$ together with the "tail" orbit $\{f^i(1), i = 0, \ldots, 2n+1\}$ whose last element maps to $\frac{7}{3}$. Thus, we can obtain patterns which force the orbit segment represented by any initial part of the f-orbit of 1 but do not force η.

A second observation is that this phenomenon is associated with tandem cycles in η. When η has no tandem cycles, then the phenomenon in 7.12 cannot occur for θ-adjusted maps of patterns.

7.13. THEOREM. *Suppose $\theta \in \mathfrak{P}$ and $\eta \in \mathfrak{S}$ has no tandem subcycles. If θ forces semi-patterns θ_m which m-shadow η for arbitrarily large m, then θ forces η.*

PROOF: Let F be θ-adjusted on $\mathcal{P}$, and let θ_m be the semi-pattern of the hypothesis which m-shadows η. Denote the degree of η by n. As in $7.5(iv)$ we can assume θ_m is a union of orbit segments of length m, and a representative $\mathcal{R}_m$ of θ_m in F consists of points $\{x_j^i(m) : i = 0, \ldots, m, j = 1, \ldots, n\}$ such that $x_j^i(m) = F^i(x_j^0(m))$ and

$$\eta^{i_1}(j_1) < \eta^{i_2}(j_2) \text{ implies } x_{j_1}^{i_1}(m) < x_{j_2}^{i_2}(m).$$

(Note that these points need not all be distinct.)

By passing to a subsequence of $\{\theta_m\}$, we can assume that for i, j fixed, the limit point

$$x_j^i = \lim_{m \to \infty} x_j^i(m)$$

exists. Note that by continuity

$$x_j^i = F^i(x_j^0) \quad j = 1, \ldots, n$$

and

$$\eta^{i_1}(j_1) < \eta^{i_1}(j_1) \text{ implies } x_{j_1}^{i_1} \leq x_{j_2}^{i_2}.$$

We will show that $x_{j_1}^{i_1} \neq x_{j_2}^{i_2}$ in the last inequality above. Let us first see how this claim implies the theorem. For $j \in \{1, \ldots, n\}$ let $k(j)$ be the length of the η-cycle through j, and set

$$I_j = [\liminf_{q \to \infty} x_j^{qk(j)}, \limsup_{q \to \infty} x_j^{qk(j)}].$$

Note that no two of these have common interior points, in fact they are disjoint unless for some j,

$$\limsup_{q \to \infty} x_j^{qk(j)} = \liminf_{q \to \infty} x_{j+1}^{qk(j+1)}.$$

In this last case, we can find a subsequence $\{\mathcal{R}_m\}$ of our sets so that the same statement is true with $x_j^{qk(j)}$ replaced by $x_j^{qk(j)}(m_i)$ (and similarly for $j+1$). But then, the semi-patterns θ_m (and their representatives $\mathcal{R}_m$) can be replaced by new semi-patterns exhibited by orbit-segments beginning at the points $x_j^{qk(j)}(m_i)$ (and for $j+1$); these satisfy the hypotheses of the theorem, but contradict the claim. Thus, we know the I_j's are disjoint.

But it is clear that I_j f-covers $I_{\eta(j)}$, and so for each j,

$$\mathcal{I}_j = \{I_j, I_{\eta(j)}, \ldots, I_{\eta^{k(j)}(j)} = I_j\}$$

is a proper loop. This gives us a periodic orbit of least period $k(j)$ with itinerary $\mathcal{I}_j$, and the union $\mathcal{R}$ of these (one per cycle in η) represents η in F. Since η is a permutation, $\mathcal{R} \cap Flat(F, \mathcal{P}) = \emptyset$ and $\theta \Rightarrow \eta$.

We are left with proving the following

CLAIM : *Suppose* $\{x_j^i(m) \mid j = 1, \ldots, n, \ i = 0, \ldots, m, \ m = 1, 2, \ldots\}$ *satisfy*

(i) $x_j^{i+1}(m) = F(x_j^i(m))$

(ii) $\eta^{i_1}(j_1) < \eta^{i_2}(j_2)$ *implies* $(\forall m)$ $x_{j_1}^{i_1}(m) < x_{j_2}^{i_2}(m)$

(iii) $x_j^i = \lim_{m\to\infty} x_j^i(m)$ *exists for all* $j \in \{1, \ldots, n\}$, $i = 0, 1, \ldots$

then

(iv) $\eta^{i_1}(j_1) < \eta^{i_2}(j_2)$ *implies* $x_{j_1}^{i_1} < x_{j_2}^{i_2}$.

Proof of claim: Of course, we need only show $x_{j_1}^{i_1} \neq x_{j_2}^{i_2}$. We will show this by contradiction.

Suppose to the contrary that $x_{j_1}^{i_1} = x_{j_2}^{i_2}$, where $\eta^{i_1}(j_1) < \eta^{i_2}(j_2)$. Note the following observation:

(*)
$$\text{if } \eta^{i_3}(j_3) \in \langle \eta^{i_1+r}(j_1), \eta^{i_2+r}(j_2)\rangle$$
$$\text{then } x_{j_3}^{i_3} = x_{j_1}^{i_1+r} = x_{j_2}^{i_2+r}.$$

In particular, taking $r = 0$, we know that unless $\eta^{i_2}(j_2) = \eta^{i_1}(j_1) + 1$, we can replace $x_{j_2}^{i_2}$ with $x_0^{i_3}$ for $j_3 = \eta^{i_1}(j_1) + 1$. Thus, we can assume

$$\eta^{i_2}(j_2) = \eta^{i_1}(j_1) + 1.$$

Now, η has no tandem cycles, so for some r,

$$|\eta^{i_1+r}(j_1) - \eta^{i_2+r}(j_2)| \geq 2,$$

and we can apply (*) again to find j_3 so that

$$j_3 \in \langle \eta^{i_1+r}(j_1), \eta^{i_2+r}(j_2)\rangle$$

and so

$$x_{j_1}^{i_1+r} = x_{j_3}^{ks} = x_{j_2}^{i_2+r}$$

where s is the period of j_3 under η and $k = 0, 1, \ldots$. In particular, we can conclude that $x_{j_3}^0$ is periodic with period (dividing) s.

Now, consider the periodic point $x_{j_3}^0 = x_{j_1}^{i_1+r} = x_{j_2}^{i_2+r}$ and let N be a common η-period of j_1, j_2 and j_3. If no image of this point is a turning point for F (in particular, if $x_{j_3}^0 \notin \mathcal{P}$), then F^k for $k = 0, \ldots, N$ is monotonic on a neighborhood U of $x_{j_3}^0$. But for m large, we have

$$x_{j_3}^k(m) \in \langle x_{j_1}^{i_1+r+k}(m), x_{j_2}^{i_2+r+k}(m)\rangle$$

for $k = 0, \ldots, N$. This in turn implies

$$\eta^k(j_3) \in \langle \eta^{i_1+r+k}(j_1), \eta^{i_2+r+k}(j_2) \rangle$$

for $k = 0, \ldots, N$. Since some value of k makes $\eta^{i_2+r+k}(j_2) = \eta^{i_2}(j_2) = \eta^{i_1}(j_1) + 1 = \eta^{i_1+r+k}(j_1)$, the open set on the right is empty, and the contradiction shows that $x_{j_3}^k$ is a turning point for some $k \in \{0, \ldots, N\}$. But in that case, we can find a neighborhood U of $x_{j_3}^0$ containing no other turning points of F^k, and such that $F[U]$ is a one-sided neighborhood of $F(x_{j_3}^1)$ mapped monotonically by F^k, $k = 0, \ldots, n$. Thus we get monotonicity of F^k, $k = 0, \ldots, n$ for the set

$$x_{j_3}^k(m),\ x_{j_1}^{i_1+r+k}(m),\ x_{j_2}^{i_2+r+k}(m),$$

and hence of η for the set

$$\eta^k(j_3),\ \eta^{i_1+r+k}(j_1),\ \eta^{i_2+r+k}(j_2),$$

which is again impossible.

This proves the claim, and hence theorem 7.13. ∎

8. TRANSITIVE PATTERNS

In this section, we use the Combinatorial Shadowing Theorem to show that in certain cases the permutation data in a pattern θ is implicit in its cycle data, in the sense that for every permutation $\eta \in \mathfrak{S}^*(\theta)$, there exists a cycle $\xi \in \mathfrak{C}(\theta)$ with $\xi \Rightarrow \eta$. This can be regarded as an extension of a theorem of Bernhardt[**Be2**].

The class of patterns which concerns us here is defined below.

8.1. DEFINITION. *Suppose $\theta \in \mathfrak{P}_n$, $n > 1$ and F is the canonical θ-adjusted map, $\mathcal{P} = \{1, \ldots, n\}$. Let $\mathcal{Z}(\theta)$ be the union of all proper loops $\mathcal{I} = \{I_0, \ldots, I_k = I_0\}$ with $I_0 = [1, 2]$. We shall call θ* **transitive** *if for each $x \in \mathcal{P}$ there exists a proper itinerary $\mathcal{I} = \{I_j\}_{j=0}^{n}$ of length $n + 1$ such that*

(i) $I_j \in \mathcal{Z}(\theta)$, $\quad j = 0, \ldots, n$

(ii) $F^j(x) \in I_j$, $\quad j = 0, \ldots, n$.

We shall call θ **strongly transitive** *if it is transitive and there is more than one prime loop with $I_0 = [1, 2]$.*

8.2. PROPOSITION. *A transitive pattern θ fails to be strongly transitive if and only if it is a permutation consisting either of a single self-tandem cycle or of a pair of tandem cycles.*

PROOF: It is easy to see that a self-tandem cycle or pair of tandem cycles forms a transitive but not strongly transitive pattern $\theta \in \mathfrak{P}_n$ (with $\mathcal{Z}(\theta) = \{I_i = \text{clos}\langle \theta^i(1), \theta^i(2) \rangle, i = 0, \ldots, n\}$.)

Conversely, suppose $\theta \in \mathfrak{P}_n$ is transitive but not strongly transitive; as in *8.1*, let F be θ-adjusted on $\mathcal{P} = \{1, \ldots, n\}$. By transitivity, at least one of any two adjacent $\mathcal{P}$-intervals belongs to $\mathcal{Z}(\theta)$, while non-strong transversality implies for any $j = 0, 1, 2, \ldots$ that each $\mathcal{P}$-interval $J \in \mathcal{Z}(\theta)$ F^j-covers precisely one element of $\mathcal{Z}(\theta)$.

Now let $A = \bigcup_{i=0}^{\infty} F^i[[1, 2]]$; this is a strictly invariant ($F(A) = A$) union of $\mathcal{P}$-intervals, and its components $A_0, \ldots, A_{k-1}$ are cyclically permuted by F. Let $[1, 2] \subset A_0 = [1, p]$, and $A_i = F^i[A_0]$. Note that $[p - 1, p] \in \mathcal{Z}(\theta)$, and $F^k[A_0] = A_0$, while $F^i[A_0] \cap A_0 = \emptyset$ unless i is a multiple of k. If $A_0 = [1, 2]$, then $A_i = \text{clos}\langle \theta^i(1), \theta^i(2) \rangle$ and our conclusion holds. Thus, we can assume $p \geq 3$.

Now, some (unique) $\mathcal{P}$-interval J_0(resp. J_1)$\in \mathcal{Z}(\theta)$ F^k-covers $[1, 2]$ (resp. $[p - 1, p]$) and necessarily $J_i \subset A_0$. Note that if some subinterval of A_0 does not belong to $\mathcal{Z}(\theta)$, it cannot F^k-cover $[1, 2]$ or $[p - 1, p]$. Thus 1 (resp. p) cannot be the image under F^k of any point of $\mathcal{P}$ interior to A_0. It follows that the set $\{1, p\}$ is f^k-invariant, so either $J_0 = [1, 2]$ or $J_0 = [p - 1, p]$. But the first case means $[p - 1, p]$ is never F^j-covered by $[1, 2]$, contradicting $[p - 1, p] \in \mathcal{Z}(\theta)$. Hence, we have $J_0 = [p - 1, p]$, and similarly $J_1 = [1, 2]$; it also follows that $F^k(p) = 1$, $F^k(1) = p$, and $[1, 2]$, $[p - 1, p]$ are the only elements of $\mathcal{Z}(\theta)$ in A_0. Thus either $p = 3$ or $p = 4$, since $[2, p - 1] \subset A_0$ is disjoint from $\mathcal{Z}(\theta)$. If $p = 4$, then $[2, 3]$ is a $\mathcal{P}$-subinterval of A_0 not in $\mathcal{Z}(\theta)$, so $F^k[2, 3] \subset [2, 3]$. Furthermore, since $[2, 3] \subset A_0$, either $[1, 2]$ or $[3, 4]$ must F^k-cover $[2, 3]$. Thus, $F^k(x) = x$ for $x = 2$ or $x = 3$. But since if $p = 4$ $F^j[J_i] \cap J_i = \emptyset$, in either case no proper itinerary has $x \in I_0 \in \mathcal{Z}(\theta)$, $F^k(x) \in I_k \in \mathcal{Z}(\theta)$, contradicting transitivity. Hence, we must have $p = 3$, and we have $F^k(1) = 3$, $F^k(2) = 2$, $F^k(3) = 1$.

Thus, the three possibilities are

(i) $A_0 = [1, 2]$ and $F^k(1) = 2$, $F^k(2) = 1$;
(ii) $A_0 = [1, 2]$ and $F^k(1) = 1$, $F^k(2) = 2$;
(iii) $A_0 = [1, 3]$ and $F^k(1) = 3$, $F^k(3) = 1$, $F^k(2) = 2$.

θ is a single self-tandem cycle in case *(i)* and a pair of tandem cycles in cases *(ii)* and *(iii)*. ∎

8.3. LEMMA. *Suppose $\theta \in \mathfrak{P}_n$ is transitive, and F is θ-adjusted on $\mathcal{P} = \{1, \ldots, n\}$. Then for every $x \in \mathcal{P}$ there exists an infinite proper itinerary $\mathcal{I}(x) = \{I_0, I_1, \ldots\}$ such that for $i = 0, 1, \ldots$ we have*

(i) $F^i(x) \in I_i$
(ii) $I_i \in \mathcal{Z}(\theta)$.

PROOF: Given $x \in \mathcal{P}$, there exists a proper itinerary $\{I_i\}_{i=0}^n$ of length $n + 1$ satisfying *(i)* and *(ii)*. Since there are only $n - 1$ $\mathcal{P}$-intervals, we have k, ℓ with $0 \le k < \ell \le n$ such that $I_k = I_\ell$.

Now (re-)define I_{k+j}, $j = \ell - k, \ell - k + 1, \ldots$ so that $I_{k+i} = I_{\ell+i}$ for $i = 0, 1, \ldots$. Clearly $\{I_i\}_{i=0}^\infty$ so defined satisfies *(ii)*. To see *(i)*, note that if $F^k(x) = F^\ell(x)$, then $F^{k+i}(x) = F^{\ell+i}(x)$ and we are done. If $F^k(x) \ne F^\ell(x)$, then $I_k = I_\ell = \text{clos}\langle F^k(x), F^\ell(x) \rangle$ is a $\mathcal{P}$-interval and $F^{\ell-k}$ interchanges its endpoints; it still follows that then $F^{k+i}(x)$, $F^{\ell+i}(x) \in I_{k+i} = I_{\ell+i}$. ∎

8.4. REMARK. *Any cycle $\theta \in \mathfrak{C}_n$, $n > 1$ is transitive. The itinerary $\mathcal{I}(x)$ described in 8.3 for $x = 1$ is then the repetition of a proper loop, sometimes called the* **fundamental loop** *of θ* [Be2].

8.5. LEMMA. *Suppose $\theta \in \mathfrak{P}_n$ is strongly transitive, and F is the canonical θ-adjusted map, $\mathcal{P} = \{1, \ldots, n\}$. Given $\varepsilon > 0$ there exists $N = N(\varepsilon)$ such that for any proper itinerary $\mathcal{I} = \{I_i\}_{i=0}^k$ with $k \ge N$ and $I_i \in \mathcal{Z}(\theta)$ for $i = 0, \ldots, k$, the representative set $\mathcal{R}(F, \mathcal{I})$ has diameter less than ε.*

PROOF: No interval in $\mathcal{Z}(\theta)$ is flat, and since θ is strongly transitive, no proper loop of intervals in $\mathcal{Z}(\theta)$ can have F an isometry on every element. Thus, for every $I \in \mathcal{Z}(\theta)$, $|DF| \ge 1$ on I, and if $J_0, \ldots, J_k = J_0$ is a proper loop with $J_i \in \mathcal{Z}(\theta)$, then $|DF| \ge 2$ on some J_i. In particular, if $\mathcal{I} = \{I_i\}_{i=0}^k$ is a proper itinerary with $I_j \in \mathcal{Z}(\theta)$ for $j = 0, \ldots, k$, then $k \ge n$ implies we must have $I_i = I_j$ for some $i < j$, so $|DF^k| \mathcal{R}(F, \mathcal{I})| \ge 2$. Inductively, we get for any integer $\ell > 0$ that $k \ge \ell n$ implies

$$|DF^k| \mathcal{R}(F, \mathcal{I})| \ge 2^\ell.$$

Now, by *1.2* (or *1.11*), F^k is affine on $\mathcal{R}(F, \mathcal{I})$, and maps this set onto I_k, which has diameter 1. Thus, if $k \ge \ell n$, we must have that the diameter of $\mathcal{R}(F, \mathcal{I})$ is at most $2^{-\ell}$. Hence any choice of $N(\varepsilon)$ with

$$N(\varepsilon) \ge -n \log_2 \varepsilon$$

works. ∎

8.6. PROPOSITION. *Given $\eta \in \mathfrak{P}_n$ strongly transitive and m a positive integer, there exists a cycle $\xi \in \mathfrak{C}^*(\eta)$ and a sub-semi-pattern $\theta \subseteq \xi$ such that θ m-shadows η. Moreover, the length of ξ can be made arbitrarily large.*

Recall that $\mathfrak{C}^*(\eta)$ denotes the set of cycles *strongly* forced by η.

PROOF: Let F be the canonical η-adjusted map, $\mathcal{P} = \{1, \ldots, n\}$. Apply *8.3* to each $\xi \in \mathcal{P}$ to obtain infinite proper itineraries $\mathcal{I}(x) = \{I_0(x), \ldots\}$ with

(i) $I_i(x) \in \mathcal{Z}(\theta)$;
(ii) $F^i(x) \in I_i(x)$.

Note also that given I, $J \in \mathcal{Z}(\theta)$ we can pick a proper itinerary

$$\mathcal{I}(I, J) = \{I_0(I, J), \ldots, I_k(I, J)\}$$

with $k = k(I, J)$ and

(i) $I_i(I, J) \in \mathcal{Z}(\theta)$;

(ii) $I_0(I, J) = I$, $I_k(I, J) = J$.

Finally, set $N = N(\frac{1}{3})$ from 8.5.

Given m, define a_ℓ, b_ℓ, c_ℓ for $\ell = 1, \ldots, n$ by

$$c_\ell = k[I_{m+n}(\ell), I_0(\ell + 1)]$$
$$b_\ell = a_\ell + m + N$$
$$a_{\ell+1} = b_\ell + c_\ell.$$

where $a_1 = 0$. Then let

$$J_i = \begin{cases} I_j(\ell) & \text{if } i = a_\ell + j \quad 0 \leq j \leq m+n \\ I_j[I_{m+n}(\ell), I_0(\ell+1)] & \text{if } i = b_\ell + j, \quad 0 \leq j \leq c_\ell \end{cases}$$

where $\ell = n + 1$ means $\ell = 1$.

Clearly, $J_i \in \mathcal{Z}(\theta)$, and $\mathcal{I}_m$ is a proper loop. Furthermore, the itinerary contains segments of length $m + N + 1$ identical to $\mathcal{I}(x)$ for each $\xi \in \mathcal{P}$. Note that we can make sure that $\mathcal{I}_m$ is a prime loop. Let $x_i = F^i(x)$, $i = 0, \ldots, a_{n+1} = p$ with $x_i \in J_i$. Then for $a_\ell \leq i \leq a_\ell + m$, and $0 \leq j \leq N$, we have $F^j(x_i)$ and $F^j(\ell)$ both in J_i, so that $|x_i - \ell| < \frac{1}{3}$ by 8.5.

Now, let ξ be the cycle represented by $\{x_i\}_{i=0}^p$ and θ the sub-semi-pattern represented by $\{x_i \mid a_\ell \leq i \leq a_\ell + m, \ell \in \mathcal{P}\}$. The disjoint sets $(\ell - \frac{1}{3}, \ell + \frac{1}{3})$ induce a block structure for θ, and since $x_i \in \theta$ for $a_\ell \leq i \leq a_\ell + m$, we have that θ m-shadows η, as required. ∎

8.7. REMARK. *The construction in the preceding proof could have been carried out replacing $\mathcal{I}(x)$, $x \in \mathcal{P}$ with any proper itinerary $\mathcal{I} = \{I_0, \ldots\}$ for which $I_i \in \mathcal{Z}(\theta)$ for all i. It follows that any point in the representative set $\mathcal{R}(F, \mathcal{I})$ of such an itinerary is the limit of a sequence of periodic points of F, with least periods growing without bound, whose itineraries also belong to $\mathcal{Z}(\theta)$.*

8.8. PROPOSITION. *A transitive but not strongly transitive pattern $\eta \in \mathfrak{P}$ has a subcycle $\xi \subseteq \eta$ which (strongly) forces every permutation strongly forced by η.*

PROOF: Let F be η-adjusted on $\mathcal{P} = \{1, \ldots, n\}$, where n is the degree of η. By 8.2, η consists of a single self-tandem cycle or a single pair of tandem cycles. In either case, the pairs $B_i = \{2i-1, 2i\}$, $i = 1, \ldots, n$ form a block structure for η over a cycle $\pi \in \mathfrak{C}_{\frac{n}{2}}$. In the first case, π is represented by the centers of the block intervals $[2i-1, 2i]$ and F has no other periodic orbits in $(2i-1, 2i)$, while in the second case $(2i-1, 2i)$ contains no periodic points. Thus, every periodic F-orbit disjoint from $\mathcal{P}_{\frac{1}{2}} = \mathcal{P} \cup \{2i - \frac{1}{2} \mid i = 1, \ldots, \frac{n}{2}\}$ is contained in the gap intervals of the block structure. It follows from lemma 3.4 that every permutation strongly forced by η is forced by π. But π is forced by either of the subcycles ξ of η. ∎

As a consequence of 8.8, 8.6 and the combinatorial shadowing theorem, we obtain the following result.

8.9. THEOREM. *Let η be a transitive pattern and n a positive integer. Then there exists a cycle $\xi \in \mathfrak{C}(\eta)$ such that $\mathfrak{S}(\xi)$ contains $\mathfrak{S}_k^*(\eta)$ for all $k \leq n$. If η*

is strongly transitive, ξ can be chosen so as to have arbitrarily large length (in particular, $\xi \in \mathfrak{C}^*(\eta)$).

PROOF: When η is transitive but not strongly transitive, the conclusion follows immediately from *8.8*.

When η is strongly transitive, we combine *8.6* and *7.10* as follows. Given n and $\pi \in \mathfrak{S}_k$ *any* permutation on $k \leq n$ elements, note that the quantities in the hypotheses of *7.10* satisfy $L \leq k \leq n$, $M \leq 2k(k-1) \leq 2(n^2 - n)$. Thus, if

$$N_n = 2(n^2 - n) + 5n - 1 = 2n^2 + 3n - 1,$$

then any semi-pattern θ which N_n-shadows η forces every permutation on $k \leq n$ elements which is strongly forced by η. In particular, by *8.6* with $m \geq N_n$, there exists $\xi \in \mathfrak{C}(\eta)$ with a sub-semi-pattern θ that N_n-shadows η. Also, by *8.6*, ξ can be chosen to be in $\mathfrak{C}_\ell(\eta)$ for ℓ arbitrarily large; in particular, for ℓ exceeding the degree of η, $\xi \in \mathfrak{C}^*(\eta)$. ∎

A simple corollary of *8.9* is the following:

8.10. THEOREM. *Let $\eta \in \mathfrak{P}_n$ be a strongly transitive pattern. Then there exists a sequence of cycles $\xi_k \in \mathfrak{C}^*(\eta)$ such that $\xi_{k+1} \Rrightarrow \xi_k$, for $k = 1, 2, \ldots$, and every permutation strongly forced by η is strongly forced by some ξ_k: that is, $\mathfrak{S}^*(\eta)$ is the increasing union of $\mathfrak{S}^*(\xi_k)$ for $k = 1, 2, \ldots$.*

PROOF: By *8.9*, there exists a sequence of cycles $\pi_n \in \mathfrak{C}^*(\eta)$ such that π_n forces any permutation of degree n or less which is strongly forced by η. The sequence $\{\xi_k\}_{k=1}^\infty$ is defined inductively as a subsequence of $\{\pi_n\}_{n=1}^\infty$. Set $\xi_1 = \pi_1$. If $\xi_k = \pi_{m_k}$ is a cycle of length ℓ_k, let

$$m_{k+1} = \max\{k, 1 + \ell_k\}$$

and set $\xi_{k+1} = \pi_{m_{k+1}}$. Then, since $\xi_k \in \mathfrak{S}_{\ell_k}^*(\eta)$ and $m_{k+1} > \ell_k$, we have $\xi - k + 1 = \pi_{m_{k+1}} \Rrightarrow \xi_k$. Also, by construction, $\mathfrak{S}_k^*(\eta) \subset \mathfrak{S}^*(\xi_k)$. ∎

Recall that by *8.4* any cycle is transitive, and by *8.2* it is strongly transitive if (and only if) it is not a doubling. More generally, *8.2* tells us that a transitive pattern without tandem subcycles is strongly transitive (the converse is false, for example $(1\ 4\ 7)(2\ 5\ 3\ 6)$ is a strongly transitive permutation with the second cycle self-tandem). For transitive *permutations* without tandem subcycles, we can use *7.13* to strengthen *8.10*. The following result restricted to cycles is due to Bernhardt [**Be2**] except for the conclusion *(ii)*, which is given by Jungreis [**J**].

8.11. THEOREM. *Suppose η is a transitive permutation with no tandem sub-cycles. Then there exists a sequence of cycles $\{\xi_k\}_{k=1}^\infty$ such that*

(i) $\eta \Rrightarrow \xi_{k+1} \Rrightarrow \xi_k$ *for $k = 1, 2, \ldots$;*
(ii) $\mathfrak{S}^*(\eta) = \bigcup_{k=1}^\infty \mathfrak{S}^*(\xi_k)$;
(iii) *any pattern which forces every ξ_k also forces η.*

PROOF: By *8.2*, our hypotheses imply η is strongly transitive. The existence of ξ_k satisfying *(i)* and *(ii)* is a special case of *8.10*. To see *(iii)*, note that by *8.6* for every m there exist cycles in $\mathfrak{C}^*(\eta)$ with sub-semi-patterns that m-shadow η; each of these is forced by some ξ_k. Thus, a pattern which forces all of the ξ_k also forces all of these semi-patterns and hence, by *7.13*, forces η. ∎

A corollary of *8.11* is the following observation.

8.12. COROLLARY. *Suppose $\eta \in \mathfrak{S}$, $\theta \in \mathfrak{S}$ are both transitive and have no tandem subcycles. Then*

(i) $\mathfrak{C}^*(\eta) = \mathfrak{C}^*(\theta)$ *implies* $\eta \Rightarrow \theta \Rightarrow \eta$;

(ii) $\mathfrak{S}^*(\eta) = \mathfrak{S}^*(\theta)$ *if and only if* $\eta = \theta$.

PROOF: By *8.11*, $\mathfrak{C}^*(\eta) \subset \mathfrak{C}^*(\theta)$ implies $\theta \Rightarrow \eta$; thus *(i)* is immediate.

To prove *(ii)*, we need only show that equivalent but unequal permutations cannot strongly force precisely the same permutations. Thus, suppose $deg(\eta) = n \le k = deg(\theta)$, $\eta \Rightarrow \theta \Rightarrow \eta$, but $\eta \ne \theta$. Let F be the canonical η-adjusted map, $\mathcal{P} = \{1, \ldots, n\}$, and take $\mathcal{Q}$ a representative of θ in F; then F is also θ-adjusted on $\mathcal{Q} \ne \mathcal{P}$. Form the set $\mathcal{R}_\ell$ of all periodic points of F with period less than or equal to ℓ, so for $\ell \ge k$, $\mathcal{Q} \cup \mathcal{P} \subset \mathcal{R}_\ell$. Consider the permutation ξ exhibited by F on $\mathcal{R}_\ell \setminus \mathcal{P}$. Any representative $\mathcal{T}$ of ξ in F is a subset of $\mathcal{R}_\ell$. But if $n < k$, then $\mathcal{T}$ cannot be a subset of $\mathcal{R}_\ell \setminus \mathcal{Q}$, so that $\xi \in \mathfrak{S}^*(\eta) \setminus \mathfrak{S}^*(\theta)$. Thus, we can assume $n = k$. Then if $\xi \in \mathfrak{S}^*(\mathcal{Q})$ we must have $\mathcal{T} = \mathcal{R}_\ell \setminus \mathcal{Q}$. Thus, there is an ordered conjugacy h_ℓ from $F|(\mathcal{R}_\ell \setminus \mathcal{P})$ to $F|(\mathcal{R}_\ell \setminus \mathcal{Q})$, for each $\ell \ge k$; as ℓ increases, each h_ℓ restricts to $h_{\ell-1}$ on $\mathcal{R}_{\ell-1} \setminus \mathcal{P}$, so setting $\mathcal{R}_\infty = \bigcup_{\ell=1}^\infty \mathcal{R}_\ell$, we get an ordered conjugacy

$$h : \mathcal{R}_\infty \setminus \mathcal{P} \to \mathcal{R}_\infty \setminus \mathcal{Q}.$$

Note that h preserves least periods, so if $\ell > k$, then h is an ordered conjugacy of the finite set $\mathcal{T}_\ell$ of all periodic points with least period ℓ to itself, hence is the identity on $\mathcal{T}_\ell$, $\ell > k$. Now, by *8.7*, $\mathcal{P} \cup \mathcal{Q}$ is contained in the closure of $\mathcal{T}_\infty = \bigcup_{\ell > k} \mathcal{T}_\ell$. Suppose $x \in \mathcal{Q} \setminus \mathcal{P}$; then x is the limit of a monotone sequence $x_i \in \mathcal{T}_\infty$; suppose for definiteness that $x_i \uparrow x$. Then $h(x) \ge \sup h(x_i) = \sup x_i = x$. Furthermore, if $h(x) > x$, then there can be no element of $\mathcal{T}_\infty$ in $\langle x, h(x) \rangle$. Similarly, if $h(x) < x$, we must have $x = \min \widetilde{x}_i$ for some sequence in $\mathcal{T}_\infty$ and $\langle x, h(x) \rangle \cap \mathcal{T}_\infty = \emptyset$. Now, $x \ne h(x)$ implies $F^i(x) \ne h(F^i(x))$ for all i, so $\langle F^i(x), h(F^i(x)) \rangle \cap \mathcal{T}_\infty = \emptyset$ for all i. Since $\mathcal{P} \cup \mathcal{Q} \subset \text{clos}\, \mathcal{T}_\infty$, this implies that x and $h(x)$ belong to tandem cycles in $\mathcal{P} \cup \mathcal{Q}$. But since the permutation $\eta \vee \theta$ exhibited by $\mathcal{P} \cup \mathcal{Q}$ is equivalent to η, we have $(\eta \vee \theta)_{**} = \eta_*$, and by definition $(\eta \vee \theta)_{**}$ must include any tandem cycles of $\eta \vee \theta$. Thus η_{**} contains the cycles through x and $h(x)$; but $\eta_{**} \subseteq \eta$, so $x \in \mathcal{P}$, contrary to assumption.

Hence, if η and θ are transitive permutations without tandem subcycles, then $\mathfrak{S}^*(\eta) = \mathfrak{S}^*(\theta)$ implies $\eta = \theta$. The opposite implication is trivial. ∎

As a special case, we note

8.13. COROLLARY. *For two cycles $\eta, \theta \in \mathfrak{C}$, the following are equivalent:*

(i) $\mathfrak{C}^*(\eta) = \mathfrak{C}^*(\theta)$

(ii) $\mathfrak{S}^*(\eta) = \mathfrak{S}^*(\theta)$

(iii) $\eta = \theta$.

9. REPRESENTATIVES OF CYCLES

Suppose $\theta \in \mathfrak{P}_n$ and $\xi \in \mathfrak{C}(\theta)$. A map f exhibiting θ may have many different periodic orbits representing ξ. Combinatorially, they may be distinguished by their placement relative to a representative of θ. The representatives of ξ in a θ-adjusted map F are those that are combinatorially forced for any map f exhibiting θ, since by *1.14* the conjugacy between F on its θ-representative $\mathcal{P}$ and f on any of its θ-representatives $\mathcal{P}'$ extends to an order-preserving conjugacy of F from the union of $\mathcal{P}$ and all the ξ-representatives in F to the union of $\mathcal{P}'$ and a collection of ξ-representatives in f, similarly situated with respect to $\mathcal{P}'$.

In this section, we explore the relation among different representatives of a given cycle ξ in a θ-adjusted map F. The picture which emerges, in particular, helps explain Baldwin's empirical observation for $\theta \in \mathfrak{C}$ [**Ba**] that very frequently the number of representatives of ξ is even.

We will concentrate on essential representatives: we will see in due course *(9.4(ii))* that non-essential representatives can occur in a θ-adjusted map only for subcycles of θ.

9.1. DEFINITION. *Given $f : I \to I$ continuous and $x \in I$ with $f(x) = x$, we call x an* **essential fixedpoint** *of f with* **index** $\sigma(f, x) = \sigma \in \{\pm 1\}$ *if for some $\varepsilon > 0$ we have*

$$\varphi_f(y) \underset{\text{def}}{=} sign[f(y) - y] = \begin{cases} -\sigma & \text{for } x - \varepsilon < y < x \\ \sigma & \text{for } x < y < x + \varepsilon. \end{cases}$$

This index is, of course, a version of the Lefschetz fixedpoint index. Note that a fixedpoint is essential if and only if it is an isolated fixedpoint interior to I and the graph of f crosses the diagonal at (x, x). The index gives the sense of the crossing. We will call x a **positive** (resp. **negative**) fixedpoint if it is essential with positive (resp. negative) index.

Essential fixedpoints are precisely the points where $f(y) - y$ switches sign. This easily gives us three observations:

9.2. REMARK. *Suppose $f : I \to I$ has finitely many fixedpoints.*

(1) *If all the essential fixedpoints of f are $\{x_1 < x_2 < \cdots < x_n\}$, then*

$$\sigma(f, x_{i+1}) = -\sigma(f, x_i) \text{ for } i = 1, \ldots, n-1.$$

(2) *If $J \subset I$ is a closed interval with $f[J] \subset \text{int } J$, then the set of essential fixedpoints of f in J is nonempty, and its leftmost and rightmost elements have negative index.*

(3) *If $J \subset I$ f-covers itself with index $\sigma(f, J, J) = \sigma$, then either one of the endpoints is fixed or int J contains an essential fixedpoint with index σ. In particular, if $\sigma(f, J, J) = -1$, then int J contains a negative fixedpoint.*

In general, the dynamics of a continuous map near an essential fixedpoint can be quite complicated. However if the map f is piecewise monotone and x is an isolated fixedpoint of f, we can understand the local dynamics quite well. Let $s = \pm 1$ and denote the one-sided neighborhoods of x by

$$\mathcal{N}(x, \varepsilon, s) = \{y \mid 0 <_s y - x <_s \varepsilon\}.$$

(Thus, $s = -1$ denotes the left side, $s = +1$ the right.) Since x is an isolated fixedpoint of f, $\varphi_f(y)$ is constant on each of $\mathcal{N}(x, \varepsilon, s)$, $s = \pm 1$, for $\varepsilon > 0$ sufficiently small. Furthermore, by piecewise-monotonicity the preimage of x in $(x - \varepsilon, x + \varepsilon)$ is a closed interval containing x, for $\varepsilon > 0$ small. Thus for $\varepsilon > 0$ small we can distinguish three kinds of dynamics on each side of x:

(i) side s is **repelling** if $y <_s f(y)$ for $y \in \mathcal{N}(x, \varepsilon, s)$;
(ii) side s is **attracting** if $x \leq_s f(y) <_s y$ for $y \in \mathcal{N}(x, \varepsilon, s)$;
(iii) side s is **flipped** onto side $-s$ if $f(y) <_s x$ for $y \in \mathcal{N}(x, \varepsilon, s)$.

Note that $\varphi_f = s$ in case *(i)* and $\varphi_f = -s$ in cases (ii) and (iii). In particular, the configurations for an essential fixedpoint are:

(a) a **repellor** if *(i)* holds for both $s = \pm 1$ (and $\varepsilon > 0$ small);
(b) an **attractor** if *(ii)* holds on one side and either *(ii)* or *(iii)* holds on the other;
(c) a **flip point** if *(iii)* holds on both sides.

Note that the name *attractor* is justified for *(b)* by the fact that x has a neighborhood, mapped into itself, in which every point has ω-limit x.

The following easy observations will be useful later.

9.3. REMARK.

(a) *If x is a repellor for f, it is a repellor for any power of f; $\sigma(f^k, x) = +1$ for all $k > 0$.*
(b) *If x is an attractor for f, it is an attractor for any power of f: $\sigma(f^k, x) = -1$ for all $k > 0$.*
(c) *If x is a flip point for f and an isolated fixedpoint for f^2, then (for f^2) (i) or (ii) holds simultaneously on both sides, so x is either a repellor or an attractor for f^2. In the first case, $\sigma(f^k, x) = (-1)^k$ for $k > 0$, while in the second $\sigma(f^k, x) = -1$ for all $k > 0$.*
(d) *A fixed turning point is either nonessential or an attractor.*

We will, in view of *(c)*, often call a flip point for f an **attracting** or **repelling flip point** (or just an attractor or repellor) depending on its nature for f^2.

9.4. LEMMA. *Suppose F is $\mathcal{P}$-adjusted and $F^k(x) = x$. Then*

(i) *If x is attracting for F^k on at least one side, there exists $z \neq x$, $z \in \mathcal{P}$ such that for $j = 0, 1, \ldots$ $F^j(x)$ and $F^j(z)$ belong to the same $\mathcal{P}$-interval;*
(ii) *if $x \notin \mathcal{P}$, then x is an essential fixedpoint of F^k; it is a repellor if $x \notin \mathcal{P}_{\frac{1}{2}}$.*
(iii) *If x is an essential fixedpoint for some power F^k of F, then it is essential for any power that fixes x.*

PROOF:

Proof of (i): We have assumed for at least one $s \in \{\pm 1\}$ and $\varepsilon > 0$ small that

$$(*) \qquad x \leq_s F^k(y) <_s y \quad \text{for all } y \in \mathcal{N}(x, \varepsilon, s).$$

If equality holds for some $y \in \mathcal{N}(x, \varepsilon, s)$ and $\varepsilon > 0$ sufficiently small, then we have $F^k[\mathcal{N}(x, \varepsilon, s)] = x$; it follows that $\mathcal{N}(x, \varepsilon, s)$ is contained in a flat $\mathcal{P}$-interval of which x is one endpoint. Letting z be the other endpoint of this interval gives us *(i)* in this case.

If $F^k(y) \neq x$ for $y \in \mathcal{N}(x, \varepsilon, s)$, then no $y \in \mathcal{N}(x, \varepsilon, s)$ has a finite orbit, and thus $\mathcal{P} \cap \mathcal{N}(x, \varepsilon, s) = \emptyset$. Now, we can increase ε as long as *(*)* holds. Let z be the other endpoint of $J = \mathcal{N}(x, \varepsilon_0, s)$, where $\varepsilon_0 = \sup\{\varepsilon \mid (*) \text{ holds}\}$. Then we must have $F^k(z) = z$, and for $j = 0, \ldots, k - 1$, $f^j[J]$ is contained in a single

$\mathcal{P}$-interval. By *1.3*, if $z \notin \mathcal{P}$ then x is an endpoint of a self-tandem $\mathcal{P}$-interval containing z. The other endpoint of this $\mathcal{P}$-interval is the point required by *(i)*.

Proof of (ii): Since $x \notin \mathcal{P}$, F^k is a homeomorphism on some neighborhood $[x-\varepsilon, x+\varepsilon]$ of x; thus F^{2k} is orientation-preserving, so that for each side $s = \pm 1$, F^{2k} is either attracting or repelling on $\mathcal{N}(x, \varepsilon, s)$. By *(i)* and *1.11*, x is a repellor unless x is at the center of a self-tandem interval. But in this case, x is a flip point for F^k, so both sides have the same character: either x is a repellor or it is an attractor; in either case it is essential.

Proof of (iii): This is an immediate corollary of *9.3*. ∎

The following two technical results will be the basis of our analysis.

9.5. LEMMA. *Suppose $f : I \to I$ is continuous and $J = [a, b] \subset I$ satisfies*
(1) *$f^i(b) \geq b$ for all i, and $f^k(b) = b$;*
(2) *$f^k(x) \leq b$ for all $x \in J$.*
Then for $i, j \in \{1, \ldots, k\}$, $f^i[J]$ and $f^j[J]$ are disjoint unless $f^i(b) = f^j(b)$.

PROOF: Suppose $f^i[J] \cap f^j[J] \neq \emptyset$, with $1 \leq i < j \leq k$. Applying f^{k-j}, we can replace j with k. Now, some $y \in J$ satisfies $f^i(y) \in f^k[J]$, so that by *(2)*, $f^i(y) \leq b$. On the other hand, $f^i(b) \geq b$, and if $f^i(b) \neq f^k(b) = b$, then some $z \neq b$ in J satisfies $f^i(z) = b$. But then $f^k(z) = f^{k-i}(b)$, and if $f^i(b) \neq b$ then *(1)* gives $f^{k-i}(b) > b$. But $f^k(z) > b$ contradicts *(2)*. ∎

9.6. PROPOSITION. *Suppose F is θ-adjusted on $\mathcal{P}$, $\theta \in \mathfrak{P}$, and suppose $p < q$ are essential fixedpoints of F^k such that*
(i) *there are no essential fixedpoints of F^k between p and q;*
(ii) *$\sigma(F^k, p) = -1$ and $\sigma(F^k, q) = +1$;*
(iii) *either $y = p$ or $y = q$ satisfies $F^i(y) > y$ for $i = 1, \ldots, k-1$.*
Then for $i = 1, \ldots, k-1$
(1) *there are no essential fixedpoints of F^i between p and q;*
(2) *for $p < x < q$, $F^i(x) \neq F^i(q)$;*
(3) *$F^i(p) \geq p$ and $F^i(q) \geq q$.*

PROOF: Since $\sigma(F^k, p) = -1$, we have $F^k(x) < x$ for x slightly to the right of p, and hence by *(i)* we have for all x, $p < x < q$
(iv) *$F^k(x) < x$ unless x is a nonessential fixedpoint of F^k (in which case $x \in \mathcal{P} \cap (p, q)$).*

Proof of (1): Suppose contrary to *(1)* that (p, q) contains an essential fixedpoint of F^i for some $i \in \{1, \ldots, k-1\}$. Pick the least i for which this occurs, and let $x \in (p, q)$ be an essential fixedpoint of F^i. We will contradict the minimality of i.

Apply the Euclidean algorithm to i and k, writing
$$k = \alpha i + \beta, \quad 0 \leq \beta < i.$$

Now $\beta \neq 0$, since otherwise x is an essential fixedpoint of F^k, contrary to *(i)*. Note that
$$F^\beta(x) = F^\beta(F^{i\alpha}(x)) = F^k(x) < x,$$
while by *(iii)* $y = p$ or q satisfies
$$F^\beta(y) > y.$$

Since $F^\beta(z) - z$ has opposite signs at x and y, the interval $\langle x, y \rangle$ contains an essential fixedpoint of F^β, contradicting minimality of i.

Proof of (2): If $F^i(x) = F^i(q)$, then $F^k(x) = F^k(q) > x$, contradicting *(iv)*.

Proof of (3): Suppose $x = p$ or $x = q$ satisfies $F^i(x) < x$ for some $i \in \{1, \ldots, k\}$. Since every point of $\mathcal{O}(x)$ is also an essential fixedpoint of F^k, we must have $F^i(x) < p$ by *(i)*. On the other hand, the other endpoint y of (p, q) satisfies $F^i(y) > y$ by *(iii)*. Thus $F^i(z) - z$ has opposite signs at $z = p$ and $z = q$, so (p, q) contains an essential fixedpoint of F^i, contrary to (1). ∎

Now, suppose $Q = \{x_1, \ldots, x_k\}$ is a representative of $\eta \in \mathfrak{C}_k$ in a piecewise-monotone map $f : I \to I$, and assume x_1, x_k are not endpoints of I. One can check that if one point in Q is an essential fixedpoint of f^k, so are the rest, and they all have the same index; in fact the one-sided behavior at each of the points in Q is the same (although side s at one point of Q may correspond to side $-s$ at another). We will call Q a **positive** (resp **negative**) **representative** of η in f if $\sigma(f^k, x) = +1$ (resp. -1) at every $x \in Q$. Note that by 9.2, if x is a positive fixedpoint of f^k and f^k has finitely many fixedpoints, then there exists a negative fixedpoint of f^k on either side of x.

The following result is the heart of our analysis in this section.

9.7. THEOREM. *Let F be θ-adjusted on $\mathcal{P}$ for $\theta \in \mathfrak{P}$ and suppose Q is a representative of the cycle $\eta \in \mathfrak{C}_k(\theta)$ in F, with x its leftmost element.*

(1) *If Q is a positive representative of η, let $y < x$ be the first essential fixedpoint of F^k to the left of x. Then y is the leftmost element of a negative representative of η.*

(2) *If Q is a negative representative of η and $k > 1$, then there exists some essential fixedpoint of F^k to the right of x. Let $y > x$ be the first such point. Then y is the leftmost point of a representative Q' of some $\pi \in \mathfrak{C}(\theta)$; either $\pi = \eta$ and Q' is a positive representative of η, or π is a 2-reduction of η and Q' is a negative representative of π.*

We note that the assumption that $k > 1$ in *(2)* is very mild. Of course, every fixedpoint of F represents the unique trivial cycle in $\mathfrak{C}_1$. Furthermore, if x is a negative fixedpoint and there exists an essential fixedpoint of F to the right of x, then the first essential fixedpoint $y > x$ is positive by 9.2 (and of course exhibits $\eta \in \mathfrak{C}_1$). Thus the only issue when $k = 1$ is the existence of an essential fixedpoint to the right of x.

PROOF:

Proof of (1): Apply 9.6 with $p = y$ and $q = x$ to conclude that y is the leftmost element of its F-orbit. Since $\sigma(F^k, y) = -1$, we have $F^k(z) \leq z \leq x$ for every $z \in [y, x]$. Thus 9.5 with $a = y$ and $b = x$ shows the intervals $F^i[y, x]$ are disjoint for $i = 1, \ldots, k$. It follows that $\mathcal{O}(y)$ also represents η.

Proof of (2): Note that since every point of Q is a negative fixedpoint of F^k, if $k > 1$ then since x is the leftmost element of Q, $F(x) > x$ is an essential fixedpoint of F^k, so there exists a first essential fixedpoint $y > x$ of F^k, which by 9.2(a) is positive, so $Q \cap [x, y] = \{x\}$. Now 9.6 applies with $p = x$, $q = y$ to give $F^i(y) \geq y$ for all i.

Note that as in the proof of *(1)*, 9.5 applies, this time with $a = x$ and $b = y$. This shows the intervals $I_j = F^j[x, y]$ are disjoint for $j = 0, \ldots, \ell - 1$, where $\ell = per(y)$. If $\ell = k$, then the F-orbits of x and y represent the same cycle. If $\ell < k$, then 9.6(2) shows that I_ℓ and I_0 can intersect only at y, so $F^\ell(x) > y$. We claim this forces $\sigma(F^\ell, y) = -1$. Otherwise, points slightly to the left of y map to the left of y while x maps to the right of y, and some interior point of (x, y) hits y under F^ℓ, contrary to 9.6(2), since $\ell < k$. Note that since $\sigma(F^\ell, y) = -1$

and $\sigma(F^k, y) = +1$, *9.3* tells us that y is a repelling flip point for F^ℓ, and k is a multiple of 2ℓ.

Now, again by *9.5* $I_{\ell+1}, \ldots, I_{2\ell-1}$ are disjoint from $I_0 \cup I_\ell$, and $I_{2\ell}$ has y as an endpoint. Suppose $2\ell < k$ and consider the location of $F^{2\ell}(x)$. Since $2\ell \in \{1, \ldots, k-1\}$, $F^{2\ell}(x) > x$ and so $F^{2\ell}(x) > y$, since it is an element of $\mathcal{Q}$ other than x. But since y must now be a positive fixedpoint of $F^{2\ell}$, points slightly to the left of y map to the left of y, and some point of (x, y) hits y under $F^{2\ell}$, contradicting *9.6(2)*. This contradiction shows that if $\ell \neq k$, then $2\ell = k$.

Thus, the sets $I_j \cup I_{j+\ell}$, $i = 0, \ldots, \ell-1$ induce a block structure on the F-orbit of x with two elements in each block, and the 2-reduction is represented by the F-orbit of y, as required. ∎

The second case of *9.7(2)* shows how a negative representative of η may be associated with a representative of a 2-reduction π of η. We formulate a kind of converse to this statement. Note that given $\pi \in \mathfrak{C}_k$, one can determine all possible doublings η of π as follows: we must have $\eta \in \mathfrak{C}_{2k}$, and the block structure $\mathcal{B} = \{B_j\}_{j=1}^k$ must have the form $B_j = \{2j-1, 2j\}$. Since $\eta \in \mathfrak{S}$, we must have $\eta(2j-1) \neq \eta(2j) \in B_{\pi(j)} = \{2\pi(j)-1, 2\pi(j)\}$. There are two bijective choices of $\eta|B_j$: the order-preserving one ($\eta(2j-1) = 2\pi(j)-1$, $\eta(2j) = 2\pi(j)$) or the order-reversing one ($\eta(2j-1) = 2\pi(j)$, $\eta(2j) = 2\pi(j)-1$). Let $\sigma_j = +1$ in the first case and $\sigma_j = -1$ in the second. A set of values $\sigma_j \in \{\pm 1\}$, $j = 1, \ldots, k$ determines a unique $\eta \in \mathfrak{S}_{2k}$ which extends π. Note that $\eta^k|B_1$ is determined by the product $\sigma = \sigma_1 \cdot \ldots \cdot \sigma_k$: if this $\sigma = +1$, we have $\eta^k(1) = 1$ and $\eta^k(2) = 2$, so η consists of two parallel copies of π: in particular, η is not a cycle. On the other hand, $\sigma = -1$ gives $\eta^k(1) = 2$, $\eta^k(2) = 1$, and $\eta \in \mathfrak{C}_{2k}$. Thus, there is a natural correspondence between the 2^{k-1} possible choices of $\sigma_1, \ldots, \sigma_k \in \{\pm 1\}$ with negative product and the cycles $\eta \in \mathfrak{C}_{2k}$ that double π. Given such a choice $(\sigma_1, \ldots, \sigma_k)$, we call the corresponding η the **doubling of π determined by** $\sigma_1, \ldots, \sigma_k$.

The relation between representatives of η and of π, where π is a 2-reduction of $\eta \in \mathfrak{C}$, is given by the following result.

9.8. PROPOSITION. *Suppose F is θ-adjusted on $\mathcal{P}$.*

(1) *If $\mathcal{Q} = \{q_1 < \cdots < q_k\}$ is a negative representative of $\pi \in \mathfrak{C}_k$ in F consisting of repellors of F^k, then the points of $\mathcal{Q}$ are positive fixedpoints for F^{2k}. Let $r_1 < q_1$ be the first essential fixedpoint of F^{2k} to the left of q_1, and let $\mathcal{R} = \mathcal{O}(r_1)$. Then $\mathcal{R} = \{r_1 < \cdots < r_{2k}\}$ represents the doubling η of π determined by $\sigma_1, \ldots, \sigma_k$, where σ_j is the sign of the slope of F at q_j.*

(2) *If $\mathcal{R} = \{r_1 < \cdots < r_{2k}\}$ is a representative of $\eta \in \mathfrak{C}_{2k}$ in F and π is a 2-reduction of η, then there exists a negative representative $\mathcal{Q} = \{q_1 < \cdots < q_k\}$ of π in F with $q_j \in I_j = [r_{2j-1}, r_{2j}]$, $j = 1, \ldots, k$.*

PROOF:

Proof of (1): A repellor with negative index for F^k has positive index for F^{2k}; thus r_1 is a negative fixedpoint of F^{2k}. Furthermore, since q_1 is a flip repellor for F^k, $F^k(f_1) \neq r_1$. Now we have the situation of *9.6* with $p = r_1$ and $q = q_1$, with k replaced by $2k$. Let $J_0 = [r_1, q_1]$ and conclude from *9.5* that the intervals $J_j = F^j[J_0]$ are disjoint for $i = 0, \ldots, k-1$ and J_{i+k} is disjoint from J_j, $0 \leq i, j < k$, unless $i = j$.

We claim $F^k(r_1) > q_1$. If $F^k(r_1) < q$, then $F^k(r_1) < r_1$, otherwise $F^k(r_1) \neq r_1$ is an essential fixedpoint of F^k between r_1 and q_1. On the other hand, since q_1 is a negative fixedpoint of F^k, we have $F^k(x) > x$ for x slightly to the left of q_1.

Hence, the open interval (r_1, q_1) contains an essential fixedpoint of F^k, hence of F^{2k}. This contradiction to the choice of r_1 proves our claim that $F^k(r_1) > q_1$.

Since $F^k(r_1) > q_1$ we can apply 9.7(2), with $x = r_1$ and $y = q_1$, to conclude that $\mathcal{O}(r_1) = \mathcal{R}$ represents a doubling η of π. Furthermore, 9.6(2) guarantees that J_j have disjoint interiors for $j = 0, \ldots, 2k - 1$. It follows that $\sigma(F, J_j, J_{j+1}) = \sigma_j$, so that η is the doubling of π determined by $\sigma_1, \ldots, \sigma_k$.

Proof of (2): By hypothesis, I_j F-covers I_{j+1}, with $\sigma(F, I_j, I_{j+1}) = \sigma_j$, $j = 1, \ldots, k$ and $\sigma_1 \cdot \ldots \cdot \sigma_k = -1$. Let $\widetilde{I}_k = I_k$ and pick $\widetilde{I}_{k-i} \subset I_{k-i}$ inductively so that $\widetilde{I}_{k-(i+1)}$ minimally F-covers $\widetilde{I}_{k-i}$ with index $\sigma_{k-(i+1)}$. Then $\widetilde{I}_0 \subset I_0 = \widetilde{I}_k$ satisfies $F^j[\widetilde{I}_0] = \widetilde{I}_j$, and $\sigma(F^k, \widetilde{I}_0, \widetilde{I}_0) = \sigma(F^k, \widetilde{I}_0, I_0) = -1$. Thus there exists an essential fixedpoint q_1 of F^k in $\widetilde{I}_0$ with index -1, but then $q_j = F^{j-1}(r_1) \in \widetilde{I}_j \subset I_j$ so $\mathcal{Q} = \mathcal{O}(q_1)$ represents π. ∎

We now combine our observations to describe the representatives of cycles.

9.9. THEOREM. *Suppose F is θ-adjusted for $\theta \in \mathfrak{P}$, and let $\eta \in \mathfrak{C}_k(\theta)$.*

(1) *F^k has an odd number of essential fixedpoints, alternating negative and positive, with negative ones at the ends. Non-essential fixedpoints of F^k must belong to blocks of tandem subcycles of θ, or be turning points of θ.*

(2) *if η is not a subcycle of θ, then every representative of η is essential.*

(3) *If*
    (a) $k > 1$,
    (b) every η-representative in F is essential (in particular, if η is not a subcycle of θ)
    (c) η has no 2-reductions,
then the number of negative η-representatives in F is nonzero and equals the number of positive η-representatives; if we order the leftmost elements of all η-representatives as $x_1 < x_2 < \cdots < x_{2\ell}$, then for $i = 1, \ldots, 2\ell$, $\sigma(F^k, x_i) = (-1)^i$.

(4) *If (a) and (b) hold but π is the (unique) 2-reduction of η and every representative of π consists of repellors (in particular if $\pi \not\subseteq \theta_{\frac{1}{2}}$), then the number of negative η-representatives is nonzero and equals the number of positive η-representatives plus the number of π-representatives whose pointwise orientation determines the doubling η. If we order the leftmost elements of all these η- and π-representatives as $x_1 < x_2 < \cdots < x_{2\ell}$, then x_i belongs to a negative η-representative for i odd and to a positive η-representative or a negative π-representative for i even.*

PROOF: *(1) follows from 9.2 and 9.4(i). (2) is 9.4(ii).*

Proof of (3): 9.7 gives a pairing of negative and positive representatives with the order described.

Proof of (4): Note that if 9.7(2) applies and y is a repellor belonging to a π-representative, then 9.8(1) gives that η is determined by the pointwise orientation of $\mathcal{O}(y)$. Again, this means we have a pairing as described. ∎

Proposition *9.8* associates to every negative representative of $\pi \in \mathfrak{C}$ outside $\mathcal{P}$ a representative of a doubling η of π. We now associate to every positive representative of π a wealth of extensions of π belonging to $\mathfrak{C}(\theta)$, provided no subcycle of θ extends π.

9.10. DEFINITION. *A pattern $\theta \in \mathfrak{P}$ positively forces $\pi \in \mathfrak{C}$ if a θ-adjusted map has a positive representative of π.*

It is clear that the condition of definition *9.10* does not depend on the choice of θ-adjusted map. Recall for §5 that $\mathfrak{h}(k, \sigma)$ denotes the k-horseshoe pattern of sign σ.

9.11. PROPOSITION. *If $\theta \in \mathfrak{P}_n$ positively forces $\pi \in \mathfrak{C}_k$ and no subcycle of θ extends π, then θ forces extensions of π by $\mathfrak{h}(2, \sigma)$ for $\sigma = -1$ and $\sigma = +1$.*

PROOF: Let $\{p_1 < p_2 < \cdots < p_k\}$ be a positive representative of π in the θ-adjusted map F. We will produce an interval $J = [q, p_1]$ intersecting a representative of an extension of π by $\mathfrak{h}(2, -1)$. Symmetric arguments give an interval $\tilde{J} = [p_k, q]$ intersecting a representative of an extension of π by $\mathfrak{h}(2, +1)$. The simplest route to obtaining $\tilde{J}$ from J is to conjugate F by a flip h, and apply our arguments (below) to $F^* = h \circ F \circ h^{-1}$, obtaining a representative of an extension of $\pi^* \in \mathfrak{C}_k$ $(\pi^*(i) = k - \pi(k - i + 1) + 1)$ by $\mathfrak{h}(2, -1)$ and noting that its image under h^{-1} gives the desired representative of an extension of π by $\mathfrak{h}(2, +1)$.

We first claim that some $b < p_1$ satisfies $F^k(b) \geq p_1$. This is because otherwise *9.5* applied to the interval $\mathcal{T} = \{x \leq p_1\}$ gives $\mathcal{T}, \ldots, F^{k-1}[\mathcal{T}]$ disjoint and $F^k(x) \leq p_1$ for all $x \in \mathcal{T}$. In particular, this would make $\mathcal{T}$ F^k-invariant, and any periodic F-orbit intersecting $\mathcal{T}$ would represent an extension of π. But the left endpoint of the ambient interval belongs to $\mathcal{P} \cap \mathcal{T}$, so its F-orbit contains a subset representing a subcycle of θ. This periodic F-orbit is contained in $\mathcal{T} \cup \cdots \cup F^{k-1}[\mathcal{T}]$ and hence intersects $\mathcal{T}$; thus it represents an extension of π, contradicting our hypothesis that no subcycle of θ extends π. This contradiction proves our claim.

Now let

$$q = \inf\{x < p_1 \mid F^k(y) < p_1 \text{ for } x < y < p_i\}.$$

By our claim, we must have $F^k(q) = p_1$. Apply *9.4* to $J = [q, p_1]$ to conclude that $F^j[J]$ are disjoint for $j = 0, \ldots, k - 1$. Since p_1 is a positive fixedpoint of F^k, it is a repellor and so does not belong to $Flat(F^k, \mathcal{P}_k)$. On the other hand, since $F^k(q) = p_1$, J cannot be contained in a $\mathcal{P}_k$-interval. Thus, J contains a point of $\mathcal{P}$, and a repetition of the argument in the preceding paragraph shows that no orbit in $\mathcal{P}$ can be contained in $J \cup \cdots \cup F^{k-1}[J]$. Thus, some $\xi \in \mathcal{P}$ satisfies $q < x < p_1$ and $F^k(x) \notin J$. By our first claim, we must have $F^k(x) < q$. Thus, we have $q < x < p_1$ with $F^k(q) = F^k(p_1) = p_1$ and $F^k(x) < q$, so by *5.2* F^k exhibits $\mathfrak{h}(2, -1)$ in J. But $F^j[J]$ are disjoint for $j = 0, \ldots, k - 1$ and $p_1 \in J$ belongs to a representative of π; hence F exhibits an extension of π by $\mathfrak{h}(2, -1)$. ∎

As a corollary of *9.11* and *6.6*, we obtain

9.12. THEOREM. *Suppose $\theta \in \mathfrak{P}$, $\pi \in \mathfrak{C}$, and no subcycle of θ extends π. If θ positively forces π (in particular, if $\pi \in \mathfrak{C}(\theta)$ is not a doubling), then θ forces simple extensions of π by all unimodal cycles.*

We close this section with two examples illustrating some of these results and their limitations.

9.13. EXAMPLE. $\theta_1 = (1\ 2\ 3\ 4)$.

This cycle, and its associated Markov graph $\mathfrak{M} = \mathfrak{M}(\theta_1)$ (together with the indices of the various coverings) is given in figure 16.

The transition matrix $A = A(\mathfrak{M})$ is

$$A = \begin{pmatrix} 0 & 1 & 0 \\ 0 & 0 & 1 \\ 1 & 1 & 1 \end{pmatrix}.$$

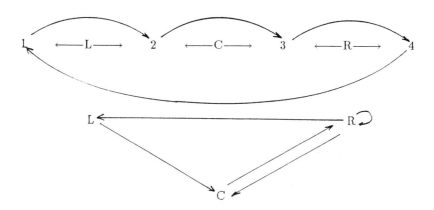

Figure 16

We will study $\mathfrak{C}_i(\theta_1)$ for $i = 2$, 3, and 6. The corresponding powers of A are

$$A^2 = \begin{pmatrix} 0 & 0 & 1 \\ 1 & 1 & 1 \\ 1 & 2 & 2 \end{pmatrix}, \quad A^3 = \begin{pmatrix} 1 & 1 & 1 \\ 1 & 2 & 2 \\ 2 & 3 & 4 \end{pmatrix}, \quad A^6 = \begin{pmatrix} 4 & 6 & 7 \\ 7 & 11 & 13 \\ 13 & 20 & 24 \end{pmatrix}.$$

Let F be θ-adjusted on $\{1, \ldots, 4\}$. The unique element of $\mathfrak{C}_2$, $\pi_1 = (1\ 2)$, has a single representative in F,

$$\mathcal{P}^* = \{p_1^* < p_2^*\}.$$

The itinerary of p_1^* is the loop

$$\mathcal{I}(p_1^* < p_2^*) = \{I_2, I_3, I_2\},$$

and $\mathcal{P}^*$ is negative. Of course, π_1 doubles the trivial permutation, represented by the (negative) fixedpoint of F between 3 and p_2^*, which is the first positive fixedpoint of F^2 to the right of p_1^*.

The unique element π_2 of $\mathfrak{C}_3(\theta_1)$ is $\pi_2 = (1\ 2\ 3)$, with two representatives in F:

$$\mathcal{P} = \{p_1 < p_2 < p_3\} \quad \text{(positive)}$$
$$\mathcal{Q} = \{q_1 < q_2 < q_3\} \quad \text{(negative)}.$$

The itineraries of the leftmost points are the loops

$$\mathcal{I}(p_1) = \{I_1, I_2, I_3, I_1\}$$
$$\mathcal{I}(q_1) = \{I_2, I_2, I_3, I_2\}$$

and the reader can check that the relative positions of points of $\mathcal{P} \cup \mathcal{Q}$ are

$$p_1 < q_1 < p_2 < q_2 < q_3 < p_3.$$

To calculate $\mathfrak{C}_6(\theta_1)$, note that the trace of A^6 is 39; subtracting the number of points of least period 1(1), 2(2), and 3(6), we are left with 30 points (or 5 orbits) of least period 6. If we denote the leftmost points of these orbits by $x_1 < x_2 < x_3 < x_4 < x_5$, we find that their itineraries are the loops

$$\mathcal{I}(x_1) = \{I_1, I_2, I_3, I_3, I_2, I_3, I_1\}$$
$$\mathcal{I}(x_2) = \{I_1, I_2, I_3, I_3, I_3, I_3, I_1\}$$
$$\mathcal{I}(x_3) = \{I_1, I_2, I_3, I_2, I_3, I_3, I_1\}$$
$$\mathcal{I}(x_4) = \{I_2, I_3, I_3, I_3, I_3, I_3, I_2\}$$
$$\mathcal{I}(x_5) = \{I_2, I_3, I_3, I_3, I_2, I_3, I_2\}.$$

One can check that $\mathcal{P}_1 = \mathcal{O}(x_1)$ and $\mathcal{Q}_1 = \mathcal{O}(x_2)$ both represent the cycle (1 2 4 5 3 6), with $\mathcal{P}_1$ negative and $\mathcal{Q}_1$ positive, and similarly $\mathcal{P}_2 = \mathcal{O}(x_4)$ (resp. $\mathcal{Q}_2 = \mathcal{O}(x_5)$) is a negative (resp. positive) representative of (1 4 3 5 2 6). In both cases, the leftmost (resp. rightmost) point of $\mathcal{P}_i$ is to the left (resp. right) of the leftmost (resp. rightmost) point of $\mathcal{Q}_i$, in agreement with 9.7. On the other hand, $\mathcal{R} = \mathcal{O}(x_3)$ is a negative representative of the cycle (1 3 5 2 4 6), which doubles $\pi_2 = (1\ 2\ 3)$. In agreement with 9.7-9.9, the (positive) fixedpoint immediately to the right of x_3 is the point p_1 in our description of $\mathfrak{C}_3(\theta_1)$.

9.14. EXAMPLE. $\theta_2 = (1\ 2\ 4\ 3)$

This cycle is $(3, +1)$-fold, with a turning point of maximum (resp. minimum) type at 2 (resp. 3). The cycle and its Markov graph are sketched in figure 17 below.

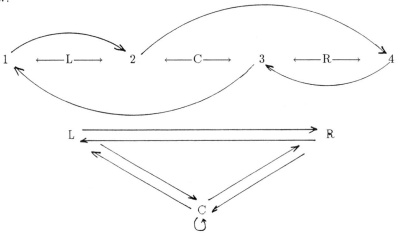

Figure 17

The transition matrix is

$$\begin{pmatrix} 0 & 1 & 1 \\ 1 & 1 & 1 \\ 1 & 1 & 0 \end{pmatrix}.$$

Again, we consider $\mathfrak{C}_i(\theta_2)$ for $i = 2$, 3 and 6, and also for $i = 5$. The corresponding powers of A are

$$A^2 = \begin{pmatrix} 2 & 2 & 1 \\ 2 & 3 & 2 \\ 1 & 2 & 2 \end{pmatrix} \quad A^3 = \begin{pmatrix} 2 & 5 & 4 \\ 5 & 7 & 5 \\ 4 & 5 & 3 \end{pmatrix}$$

$$A^5 = \begin{pmatrix} 20 & 29 & 21 \\ 29 & 41 & 29 \\ 21 & 29 & 20 \end{pmatrix} \quad A^6 = \begin{pmatrix} 50 & 70 & 49 \\ 70 & 99 & 70 \\ 49 & 70 & 50 \end{pmatrix}.$$

Again, let F be θ-adjusted on $\{1, \ldots, 4\}$.

The unique element of $\mathfrak{C}_2$, $\pi_1 = (1\ 2)$, has three representatives in F,

$$\mathcal{P}^* = \{p_1^* < p_2^*\}$$
$$\mathcal{Q}^* = \{q_1^* < q_2^*\}$$
$$\mathcal{R}^* = \{r_1^* < r_2^*\}$$

corresponding to the loops

$$\mathcal{I}(p_1^*) = \{I_1, I_2, I_1\}$$
$$\mathcal{I}(q_1^*) = \{I_1, I_3, I_1\}$$
$$\mathcal{I}(r_1^*) = \{I_2, I_3, I_2\}.$$

If we denote the (unique) fixedpoint of F by s, one can check that the order of $\mathcal{P}^* \cup \mathcal{Q}^* \cup \mathcal{R}^* \cup \{s\}$ is

$$p_1^* < q_1^* < r_1^* < s < p_2^* < q_2^* < r_2^*.$$

$\mathcal{P}^*$ and $\mathcal{R}^*$ are both negative representatives of π_1, while $\mathcal{Q}^*$ is the positive one. Of course, π_1 doubles the trivial permutation, reflected in the fact that the first positive fixedpoint of f^2 to the right of r_1^* is s.

Both elements of $\mathfrak{C}_3$, $\pi_2 = (1\ 2\ 3)$ and $\tilde{\pi}_2 = (1\ 3\ 2)$, have two representatives in F. The representatives of π_2 are $\mathcal{P} = \{p_1 < p_2 < p_3\}$ (negative), and $\mathcal{Q} = \{q_1 < q_2 < q_3\}$ (positive), where

$$\mathcal{I}(p_1) = \{I_1, I_2, I_3, I_1\}$$
$$\mathcal{I}(q_1) = \{I_1, I_2, I_2, I_1\},$$

and it can be checked that

$$p_1 < q_1 < p_2 < q_2 < q_3 < p_3.$$

An analogous situation holds for representatives of $\tilde{\pi}_2$.

A complete description of $\mathfrak{C}_6(\theta)$ is a bit unwieldy. The trace of A^6 is 199, and even after subtracting the (1) fixedpoint, (6) period 2 points, and (12) period 3 points, we are left with 180 points of period 6, corresponding to 30 orbits. We will concentrate here on doublings of $\pi_2 = (1\ 2\ 3)$. There is only one negative representative $\mathcal{P}$ of π_2 in F, and its orientation determines the doubling

$$\eta = (1\ 3\ 6\ 2\ 4\ 5).$$

Our results predict that F will exhibit a unique (negative) representative of η. Note that η has a turning point of maximum (resp. minimum) type at 3 (resp. 5); using this, we can verify that the only representative of η in F is the orbit whose leftmost point has as its itinerary the loop

$$\{I_1, I_2, I_3, I_1, I_2, I_2, I_1\}.$$

If we compare doublings of $\pi_2 = (1\ 2\ 3)$ in this example and the previous one, we see that both of the cycles θ_1 and θ_2 force the common cycle π_2, but they force no common extensions of π_2, even though each of θ_1 and θ_2 individually forces infinitely many extensions of π_2.

Finally, we consider $\mathfrak{C}_5(\theta_2)$. There are 16 orbits of least period 5, and we shall not attempt to list them here. However, we note that there are precisely four representatives of $\pi = (1\ 3\ 4\ 2\ 5)$, which we label $\widetilde{\mathcal{P}}$, $\widetilde{\mathcal{Q}}$, $\widetilde{\mathcal{R}}$, and $\widetilde{\mathcal{T}}$, denoting their leftmost points $\widetilde{p}_1$, $\widetilde{q}_1$, $\widetilde{r}_1$, $\widetilde{t}_1$. Their itineraries are the loops listed below:

$$\mathcal{I}(\widetilde{p}_1) = \{I_1, I_2, I_2, I_1, I_2, I_1\}$$
$$\mathcal{I}(\widetilde{q}_1) = \{I_1, I_2, I_2, I_1, I_3, I_1\}$$
$$\mathcal{I}(\widetilde{r}_1) = \{I_1, I_2, I_2, I_2, I_3, I_1\}$$
$$\mathcal{I}(\widetilde{t}_1) = \{I_1, I_2, I_2, I_2, I_2, I_1\}.$$

One can check that $\widetilde{\mathcal{P}}$ and $\widetilde{\mathcal{R}}$ are negative and $\widetilde{\mathcal{Q}}$ and $\widetilde{\mathcal{T}}$ are positive representatives of π, and that, in agreement with *9.9*, their order is

$$\widetilde{p}_1 < \widetilde{q}_1 < \widetilde{r}_1 < \widetilde{t}_1.$$

On the other hand, the *right*most points of these orbits have the order

$$\widetilde{t}_5 < \widetilde{p}_5 < \widetilde{q}_5 < \widetilde{r}_5$$

which also agrees with *9.9*. But note that the alternation of rightmost points pairs $\widetilde{\mathcal{P}}$ with $\widetilde{\mathcal{T}}$ and $\widetilde{\mathcal{Q}}$ with $\widetilde{\mathcal{R}}$, which is different from the pairing via alternation of leftmost points, namely $\widetilde{\mathcal{P}}$ with $\widetilde{\mathcal{Q}}$ and $\widetilde{\mathcal{R}}$ with $\widetilde{\mathcal{T}}$.

10. FORCING AND DEGREE

In §5 we considered the forcing relation among (k, σ)-fold patterns for $k \geq 2$ and $\sigma = \pm 1$ fixed. Proposition 5.4 and Theorem 5.5 together show that $\mathfrak{h}(k, \sigma)$ is the only (k, σ)-fold pattern without (flat or) tandem cycles which is forced by no other (k, σ)-fold pattern. Using 5.4(iii) and 8.9, we see that every (k, σ)-fold cycle (resp., pattern without tandem cycles) is forced by some other (k, σ)-fold cycle, so that the cycles or permutations of a given fold type cannot in general be expected to contain maximal elements with respect to forcing.

Baldwin [**Ba**] posed a similar question, for cycles, with fold type replaced by degree: which cycles $\theta \in \mathfrak{C}_n$, n fixed, are forced by no other cycles of the same degree? Baldwin's question has been answered by Jungreis [**J**]. In this section, we take up the natural analogues of Baldwin's question for permutations and patterns of fixed degree, and place Jungreis' description in a broader context.

The k-horseshoes are natural candidates for forcing-maximal elements of $\mathfrak{P}_{k+1}$, $k \geq 2$. That they are indeed maximal is a corollary of the following results.

10.1. LEMMA. *If $\theta \in \mathfrak{P}_{k+1}$ ($k \geq 2$) is not forced by any k-horseshoe, then for each $i = 1, \ldots, k$, one of the following holds:*

(a) $\theta(i) = \theta(i + 1)$
(b) $\theta(i) = i$ *and* $\theta(i + 1) = i + 1$
(c) $\theta(i) = i + 1$ *and* $\theta(i + 1) = i$.

PROOF: Suppose for some $i \in \{1, \ldots, k\}$, none of *(a)-(c)* holds; in particular, $\theta(i) <_\sigma \theta(i+1)$ for some $\sigma \in \{\pm 1\}$. Let $\mathcal{P}$ consist of all integers and half-integers in $[\frac{1}{2}, k + \frac{3}{2}]$ except $i + \frac{1}{2}$. Define $f : \mathcal{P} \to \mathcal{P}$ as follows: for $\ell = 1, \ldots, k + 1$,

$$f(\ell) = \theta(\ell) \in \mathcal{P};$$

for $\ell = j + \frac{1}{2}$, $j \neq i$,

$$f(\ell) = \begin{cases} \frac{1}{2}, & \text{if } (-1)^{i-j}\sigma = \text{ sign } (i - j); \\ k + \frac{3}{2}, & \text{if } (-1)^{i-j}\sigma = \text{ sign } (j - i). \end{cases}$$

Note that

$$(*) \qquad f(i - \frac{1}{2}) <_\sigma f(i) <_\sigma f(i + 1) <_\sigma f(i + \frac{3}{2}),$$

and the restriction of f to the half-integers in $\mathcal{P}$ exhibits a k-horseshoe, while its restriction to the integers exhibits θ. Now the pair $\{\frac{1}{2}, 1\}$ and/or $\{k + 1, k + \frac{3}{2}\}$ belongs to a tandem block if and only if one of the following holds:

(i) $f(\frac{1}{2}) = \frac{1}{2}$ and $f(1) = 1$;
(ii) $f(k + 1) = k + 1$ and $f(k + \frac{3}{2}) = k + \frac{3}{2}$;
(iii) $f(\frac{1}{2}) = k + \frac{3}{2}$, $f(1) = k + 1$, $f(k + 1) = 1$, $f(k + \frac{3}{2}) = \frac{1}{2}$.

We modify $\mathcal{P}$ by deleting 1 in the first case, $k + 1$ in the second and both 1 and $k + 1$ in the third; otherwise, we leave $\mathcal{P}$ alone. Let $\mathcal{Q} \subset \mathcal{P}$ denote the integers in $\mathcal{P}$, together with $\frac{1}{2}$ (in case *(i)* or *(iii)*) and $k + \frac{3}{2}$ (in case *(ii)* or *(iii)*), and $\mathcal{P}_* \subset \mathcal{P}$ the half-integers.

Let the patterns exhibited by f on $\mathcal{P}$, $\mathcal{Q}$, and $\mathcal{P}_*$ be denoted η, θ, and π (resp.). The use of θ for $f|\mathcal{Q}$ is justified by the definition of $f(\ell)$ for ℓ an integer. Note that π is a k-horseshoe. We claim that $\pi = \eta_{**}$, the essential subpattern of η as defined in 2.6. Note first that every $\mathcal{P}_*$-interval except $[i - \frac{1}{2}, i + \frac{3}{2}]$ contains at most one integer, and since the endpoints of each map to $\frac{1}{2}$ and $k + \frac{3}{2}$, which are beyond the minimum and maximum integers in $\mathcal{Q}$, f is strictly monotone on each such $\mathcal{P}_*$-interval. On the other hand, the inequality *(*)* shows strict monotonicity of f on $[i - \frac{1}{2}, i + \frac{3}{2}]$. Thus, f has no flat intervals, and every turning point belongs to $\mathcal{P}_*$. Finally, there are no tandem blocks. No element of $\mathcal{P}_*$ can belong to a tandem block, because a periodic element of $\mathcal{P}_*$ must be either $\frac{1}{2}$ or $k + \frac{3}{2}$, and our deletions in cases *(i)-(iii)* made sure that neither of these belongs to a tandem block. On the other hand, the only pair of adjacent elements of $\mathcal{P}$ which is disjoint from $\mathcal{P}_*$ is $i, i + 1$, and since *(b)* and *(c)* do not hold, they cannot be tandem, either.

Thus, by 2.6, π is the essential subpattern of η; in particular, $\pi \Rightarrow \eta \Rightarrow \theta$, as desired. ∎

It is easy to see that any pattern $\theta \in \mathfrak{P}_{k+1}$ $(k \geq 2)$ which satisfies the conditions of 10.1 must be monotone, either **nondecreasing** $(\theta(i) \leq \theta(i + 1)$ for all $i)$ or **nonincreasing** $(\theta(i) \geq \theta(i + 1)$ for all $i)$. Thus, 10.1 says that the k-horseshoes force all non-monotone patterns of degree $k + 1$; with a few "silly" exceptions, the k-horseshoes are the only forcing-maximal elements of $\mathfrak{P}_{k+1}$. The exceptions are defined below.

10.2. DEFINITION. Given $1 \leq i_1 \leq i_2 \leq k + 1$, define $\psi = \phi_{k+1}(i_1, i_2) \in \Pi_{k+1}$ by (figure 18)

$$\psi(i) = \begin{cases} i_1 & \text{if } i \leq i_1 \\ i & \text{if } i_1 \leq i \leq i_2 \\ i_2 & \text{if } i_2 \leq i. \end{cases}$$

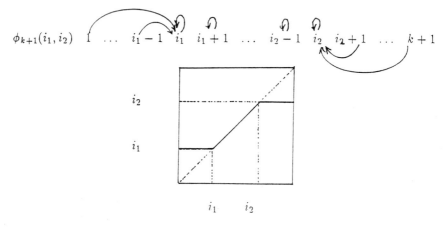

Figure 18

10.3. DEFINITION. Given $1 \leq j \leq k$, define $\psi = \tilde{\phi}_{k+1}(j) \in \Pi_{k+1}$ by (figure 19)

$$\psi(i) = \begin{cases} j + 1 & \text{if } i \leq j \\ j & \text{if } i \geq j + 1. \end{cases}$$

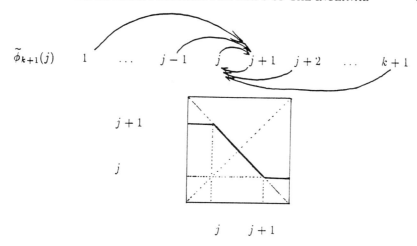

Figure 19

The following shows that the patterns defined above are the only non-horseshoe candidates for maximal elements of $\mathfrak{P}_{k+1}$. We will refer to these collectively as the **silly patterns** .

10.4. LEMMA. *If $\theta \in \mathfrak{P}_{k+1}$ $(k \geq 2)$ is not forced by any k-horseshoe, then θ is silly.*

PROOF: By hypothesis, every pair $i, i+1$, $i = 1, \ldots, k$ satisfies one of *(a)-(c)* of *10.1*. Suppose first that some pair $i, i+1$ satisfies *(c)*. Then it is easy to see (by induction) that no pair $\ell, \ell+1$ with $\ell+1 \leq i$ or $\ell \geq i+1$ can satisfy *(b)* or *(c)*; hence, $\theta = \tilde{\phi}_{k+1}(j)$ with $j = 1$.

On the other hand, if *(c)* does not hold for any pair, then each endpoint of any maximal flat block of θ must be either 1, $k+1$, or a fixedpoint. Since it is impossible for both endpoints of a flat block to be fixed, it follows easily that $\theta = \phi_{k+1}(i_1, i_2)$ for some $1 \leq i_1 \leq i_2 \leq k+1$. ∎

Finally, the last step in determining the maximal elements of $\mathfrak{P}_{k+1}$ is the following.

10.5. LEMMA. *If θ and η are silly patterns of the same degree then $\theta \Rightarrow \eta$ implies $\theta = \eta$.*

PROOF: Let $\deg \theta = \deg \eta = k+1$. Note that the case $k = 0$ is trivial, so we can assume $k \geq 1$. Suppose $\theta \neq \eta$. We will construct $f : [1, k+1] \to [1, k+1]$ which exhibits θ on $\mathcal{P} = \{1, \ldots, k+1\}$ but does not exhibit η.

Let F be the canonical θ-adjusted map; we define f on each interval $[i, i+1]$, $i = 1, \ldots, k$ by

$$f(x) = F(x) + \varepsilon(i)[(x - i) - (x - i)^2],$$

where $\varepsilon(i)$ is defined by the following:
Case 1: $\theta = \phi_{k+1}(i_1, i_2)$:

   if $i_1 < i_2$, then

$$\varepsilon(i) = \begin{cases} +1 & \text{for } i < i_1; \\ 0 & \text{for } i_1 \leq i < i_2; \\ -1 & \text{for } i_2 \leq i; \end{cases}$$

if $i_1 = i_2 \leq k$, then $\varepsilon(i) = +1$ for all i;
if $i_1 = i_2 = k + 1$, then $\varepsilon(i) = -1$ for all i;

Case 2: $\theta = \widetilde{\phi}_{k+1}(j)$:

$$\varepsilon(i) = \begin{cases} -1 & \text{if } i < j; \\ 0 & \text{if } i = j; \\ +1 & \text{if } i > j. \end{cases}$$

The reader can check that every preperiodic point of f belongs to $\mathcal{P}$, except for the center of the tandem block when $\theta = \widetilde{\phi}_{k+1}(j)$. It is then clear that there is no representative of η in f. ∎

The characterization of maximal patterns of fixed degree is then given as follows.

10.6. THEOREM. *For $k \geq 0$, $\theta \in \mathfrak{P}_{k+1}$ is not forced by any $\eta \in \mathfrak{P}_{k+1}$ with $\eta \neq \theta$ if and only if θ is either a k-horseshoe ($k \geq 2$) or a silly pattern.*

PROOF: First, note that for $k = 0, 1$, every element of $\mathfrak{P}_{k+1}$ is silly, and in these cases $\eta, \theta \in \mathfrak{P}_{k+1}$, $\eta \Rightarrow \theta$ implies $\eta = \theta$ (for $k = 1$, use *10.5*).

Suppose then that $k \geq 2$. If θ is not forced by any other element of $\mathfrak{P}_{k+1}$, then by lemma *10.4* θ is a horseshoe or a silly pattern.

That horseshoes are not forced by any other patterns of the same degree follows from *10.4* and fold type considerations. Thus, in view of *10.5.* to complete the proof we need only show that no k-horseshoe can force a silly pattern θ of degree $k + 1$. Let f be $\mathfrak{h}(k, \sigma)$-linear on $\mathcal{P} = \{1, \ldots, k + 1\}$ and suppose $\mathcal{Q}$ is a representative of θ in f. Since f is expanding on each $\mathcal{P}$-interval, a flat or tandem block in $\mathcal{Q}$ can intersect each $\mathcal{P}$-interval in at most one point. Since every pair of adjacent points in $\mathcal{Q}$ is flat or tandem, the k $\mathcal{P}$-intervals must separate the $k + 1$ points of $\mathcal{Q}$, a contradiction. ∎

Next, we take up the question of which permutations are maximal for the forcing relation restricted to permutations of fixed degree. In *10.6*, we saw that the permutation $\phi_n(1, n)$ consisting of n fixedpoints is maximal in $\mathfrak{S}_n$. The following results will give us a class of other maximal elements of $\mathfrak{S}_n$, with characteristics that are in many ways opposite to those of $\phi_n(1, n)$.

As in §2, we will refer to endpoints and turning points as **critical points** of $\theta \in \mathfrak{S}_n$. Also, the set of **neighbors** of $i \in \{1, \ldots, n\}$ is defined by

$$\mathcal{N}(i) = \{j \mid j \in \{1, \ldots, n\} \text{ and } |i - j| \leq 1\}.$$

10.7. LEMMA. *Suppose the permutation $\theta \in \mathfrak{S}_n$, $n \geq 3$, has a critical point whose image is not critical. Then there exists a permutation $\eta \neq \theta$, of the same degree and with the same number of cycles as θ, with $\eta \Rightarrow \theta$.*

PROOF: By assumption, some $t \in \{1, \ldots, n\}$ satisfies, for some $\sigma = \pm 1$,

(a) $\theta(t) \geq_\sigma \theta[\mathcal{N}(t)]$
(b) $\theta(t) = u \in \{2, \ldots, n - 1\}$, and $\theta(u) \in \langle \theta(u - 1), \theta(u + 1) \rangle$.

Let $\mathcal{P} = \{1, \ldots, n\}$, and form $\widetilde{\mathcal{P}}$ by deleting u and inserting $\widetilde{u} = u + \frac{3}{2}\sigma$. Thus, $\widetilde{\mathcal{P}}$ has n points; the order-preserving bijection $h : \mathcal{P} \to \widetilde{\mathcal{P}}$ is given by

$$h(i) = \begin{cases} i & \text{if } i <_\sigma u \text{ or } i >_\sigma u + \sigma \\ u + \sigma & \text{if } i = u \\ \widetilde{u} & \text{if } i = u + \sigma. \end{cases}$$

Now, define $f : \widetilde{\mathcal{P}} \to \widetilde{\mathcal{P}}$ by

$$f(i) = \begin{cases} \widetilde{u} & \text{if } i = t \\ v = \theta(u) & \text{if } i = \widetilde{u} \\ \theta(i) & \text{otherwise.} \end{cases}$$

Then f exhibits the permutation $\eta = h^{-1} \circ f \circ h$ on $\widetilde{\mathcal{P}}$. Clearly, η and θ have the same degree and the same number of cycles. We need to show that $\eta \Rightarrow \theta$. If we let $\widetilde{\theta} \subseteq \theta$ be the union of subcycles of θ which do not include t, then $\widetilde{\theta} \subseteq \eta$; it is exhibited by f on a subset $\widetilde{\mathcal{Q}}$ of $\mathcal{P} \cup \widetilde{\mathcal{P}} \setminus \{t\} = \mathcal{P} \setminus \{u, t\}$.

Now, extend f to F, η-adjusted on $\widetilde{\mathcal{P}}$, and let G be θ-adjusted on $\mathcal{P}$. We assign to the subcycle of θ through t (and u) two proper itineraries $\mathcal{I}_\pm = \{I_0^\pm, \ldots, I_k^\pm\}$ as follows: set

$$I_0^\pm = \operatorname{clos}\langle u \pm \sigma, u \rangle$$

and define $I_{j+1}^\pm$ inductively to be the (unique) $\mathcal{P}$-interval which is G-covered by $I_j^\pm$ and contains $G^j(u)$. If $k = per(u)$, then since $I_{k-1}^\pm$ both contain t, which is critical,

$$I_k^+ = I_k^- = I_0^-.$$

We use $\mathcal{I}_+$ to define a loop $\widetilde{\mathcal{I}}_+ = \{\widetilde{I}_0^+, \ldots, \widetilde{I}_k^+\}$, as follows.

(i) If I_j^+ contains u, set $\widetilde{I}_j^+ = \operatorname{clos}\langle u - \sigma, u + \sigma \rangle$. Note that then $\widetilde{I}_j^+$ F-covers I_{j+1}^+, with the same orientation as the G-cover of I_{j+1}^+ by I_j^+.

(ii) If

$$I_j^+ = \operatorname{clos}\langle u + \sigma, u + 2\sigma \rangle,$$

then one of the $\widetilde{\mathcal{P}}$-intervals contained in I_j^+, $\operatorname{clos}\langle u + \sigma, \widetilde{u} \rangle$ or $\operatorname{clos}\langle \widetilde{u}, u + 2\sigma \rangle$, F-covers I_{j+1}^+, with the same orientation as the G-cover of I_{j+1}^+ by I_j^+; pick this $\widetilde{\mathcal{P}}$-interval as $\widetilde{I}_j^+$.

(iii) If I_j^+ contains neither u nor $\widetilde{u}$, then I_j^+ is also a $\widetilde{\mathcal{P}}$-interval, and $F|I_j^+ = G|I_j^+$; we take $\widetilde{I}_j^+ = I_j^+$ in this case.

Thus, $\widetilde{\mathcal{I}}$ is a proper itinerary for F. Furthermore, the definition gives

$$\widetilde{I}_0 = \widetilde{I}_k = \operatorname{clos}\langle u - \sigma, u + \sigma \rangle,$$

so it is a loop. Note, finally, that $I_i^\pm \leq I_j^\pm$ implies $\widetilde{I}_i^\pm \leq \widetilde{I}_j^\pm$.

The same procedure applied to $\mathcal{I}_-$ gives a proper loop $\widetilde{\mathcal{I}}_-$ for F. Note that $\widetilde{I}_1^+ \neq \widetilde{I}_1^-$, because the monotonicity of G on $\operatorname{clos}\langle u - \sigma, u + \sigma \rangle$ guarantees that I_0^+ and I_0^- G-cover distinct $\mathcal{P}$-intervals. Note also that for $j = 0, \ldots, k$, $G^j[\mathcal{R}(G, \mathcal{I}_+)]$ and $G^j[\mathcal{R}(G, \mathcal{I}_-)]$ meet at $G^j(u)$, so that $\widetilde{\mathcal{Q}}$ is disjoint from these sets, and hence from $F^j[\mathcal{R}(F, \mathcal{I}_+)] \cup F^j[\mathcal{R}(F, \mathcal{I}_-)]$.

Finally, we claim $\widetilde{\mathcal{I}}_+$ is a prime loop. Suppose $j > 0$ is the least period of $\widetilde{\mathcal{I}}_+$, so $\widetilde{I}_{i+j}^+ = \widetilde{I}_i^+$ for $i = 0, \ldots, k - j$. In particular, $\widetilde{I}_j^+ = \widetilde{I}_0^+$, so $I_j^+ = I_0^\pm$, and if $j < k$, then $G^j(u) = u \pm \sigma$. Since $\widetilde{I}_{i+j}^+ = \widetilde{I}_i^+$ for $i = 1, \ldots, j - 1$, we see that $G^j[I_i^\pm] = I_i^\pm$, and u and $u \pm \sigma$ belong to tandem cycles. But $u + \sigma \notin I_k^\pm$, so we must have $u, u - \sigma$ tandem, and $I_j^+ = G^j[I_0^+] = I_0^-$. But then $I_{j+1}^+ \neq I_1^+$, contradicting the choice of j.

Now, let $x \in \mathcal{R}(F, \widetilde{\mathcal{I}}^+)$ be F-periodic with period k. Then $\mathcal{O}_F(x)$ represents the same cycle as $\mathcal{O}_G(u)$. Furthermore, $G^j(u)$ and $F^j(x)$ belong to the same $\mathcal{P}$-interval, except if one belongs to I_0^+ and the other to I_0^-. Thus, they belong to the same $\widetilde{Q}$-interval. We note that $x \neq u - \sigma$, since $x \in I_0^+$; hence the union Q of $\widetilde{Q}$ with $\mathcal{O}_F(x)$ is a representative of θ in F. It follows that $\eta \Rightarrow \theta$, as required. ∎

As a corollary of *10.7*, we obtain the following necessary conditions for θ to be maximal in $\mathfrak{S}_n$.

10.8. PROPOSITION. *Suppose* $\theta \in \mathfrak{S}_n$ *is maximal in* $\mathfrak{S}_n$. *Then* $\theta = \theta_{**}$, *and the subcycles of* θ *fall into two mutually disjoint classes:*

    (i) **critical cycles**, *in which every element is a critical point for* θ;
    (ii) **non-critical cycles**, *which contain no critical points for* θ.

Every non-critical point belongs to a tandem block of θ.

PROOF: By *10.7*, every cycle is either critical or non-critical (and not both). Note also that θ is essential; otherwise $\theta_{**} \in \mathfrak{S}_k$, $k < n$ and $\theta_{**} \Rightarrow \theta$. By adding $n - k$ fixedpoints, we create $\eta \in \mathfrak{S}_n$, $\eta \neq \theta$, with $\eta \Rightarrow \theta$. It follows from *2.4* that non-critical cycles are tandem to other cycles. ∎

A further necessary condition for maximality is given by the next result.

10.9. LEMMA. *If* $\theta \in \mathfrak{S}_n$ *and* $x, y \in \mathcal{P} = \{1, \ldots, n\}$ *satisfy*

$$\theta[\mathcal{N}(x)] \leq \theta(x) < \theta(y) \leq \theta[\mathcal{N}(y)],$$

then there exists $\eta \in \mathfrak{S}_n$, $\eta \neq \theta$ *with* $\eta \Rightarrow \theta$.

PROOF: Define $\eta \in \mathfrak{S}_n$ by

$$\eta(i) = \begin{cases} \theta(i) & \text{if } i \neq x, y \\ \theta(x) & \text{if } i = y \\ \theta(y) & \text{if } i = x. \end{cases}$$

Let G (resp. F) be θ-adjusted (resp. η-adjusted) on $\mathcal{P} = \{1, \ldots, n\}$. Then $F = G$ except on the convex hulls of $\mathcal{N}(x)$ and $\mathcal{N}(y)$. Each of these consists of (at most) two $\mathcal{P}$-intervals, $[x - 1, x]$, $[x, x + 1]$ (resp. $[y - 1, y]$, $[y, y + 1]$), and for I any of these intervals, $F[I] \supset G[I]$. Thus, any proper loop for G is also a proper loop for F and the orientation of F and G on each interval is the same. Of course, every subcycle of θ which includes neither x nor y is also a subcycle of η, and the union of these forms a subpermutation $\widetilde{\theta}$ of θ which is also a subpermutation of η, exhibited by F on $\widetilde{Q} \subset \mathcal{P}$. It remains to show that η forces the subcycle(s) of θ through x and y. We concentrate on x; the corresponding argument for y will be clear. As in *10.7*, we attach to $u = \theta(x)$ two itineraries for G, $\mathcal{I}_{\pm} = \{I_0^{\pm}, \ldots, I_k^{\pm}\}$ by setting $I_0^{\pm} = \text{clos}\langle u \pm 1, u\rangle$ and $I_{j+1}^{\pm}$ the (unique) $\mathcal{P}$-interval G-covered by $I_j^{\pm}$ which contains $G^{j+1}(u)$. If $k = per(u)$, then as before $I_k^+ = I_k^- = I_0^-$. Now, note that $I_{k-1}^{\pm}$ F-covers $I_0^{\pm} = [u, u + 1]$ as well as $I_0^- = [u - 1, u]$. Thus, the itinerary $\widetilde{\mathcal{I}}_+ = \{\widetilde{I}_0^+, \ldots, \widetilde{I}_k^+ = \widetilde{I}_0^+\}$, where $\widetilde{I}_i^+ = I_i^+$ for $i = 0, \ldots, k - 1$, is also a proper loop for F, and clearly $\widetilde{\mathcal{I}}_+ \neq \mathcal{I}_-$. Let $\widetilde{u} \in \mathcal{R}(F, \widetilde{\mathcal{I}}_+)$ be k-periodic for F. If $\widetilde{\mathcal{I}}_+$ is prime, then $\mathcal{O}_F(\widetilde{u})$ represents the subcycle of θ through u (and x), and $u - 1 \notin \mathcal{O}_F(\widetilde{u})$. But if $\widetilde{\mathcal{I}}_+$ is not prime, then u and $u + 1$ belong to the same G-orbit, since $I_j^+ = I_0^+$ for some $j < k$, and since $G[I_{k-1}] \leq u$, it follows that u and $u + 1$ belong to a self-tandem subcycle of θ, which contradicts the fact that x is a turning point.

But then, as in *10.7*, the union Q of $\widetilde{Q}$ with $\mathcal{O}_F(\widetilde{u})$ and the corresponding F-orbit which exhibits the subcycle of θ through y is a representative of θ in F, so $\eta \Rightarrow \theta$. ∎

10.8 and *10.9* give us necessary conditions for a permutation $\theta \in \mathfrak{S}_n$ to be maximal. These conditions, however, put no restrictions on the tandem blocks of a permutation. The following "flip" permutations will help us characterize such tandem blocks.

10.10. DEFINITION. *The 2n-flip* $\pi_{2n} \in \mathfrak{S}_{2n}$ *is defined by (see figure 20)*

$$\pi_{2n}(i) = 2n + 1 - i \quad i = 1, \ldots, 2n.$$

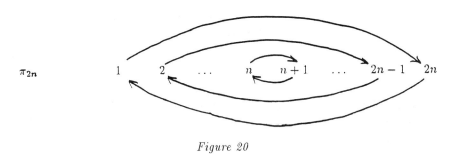

Figure 20

10.11. LEMMA. *For* $n \geq 3$, *there exists* $\theta_{2n} \in \mathfrak{S}_{2n}$, $\theta_{2n} \neq \pi_{2n}$, *with* $\theta_{2n} \Rightarrow \pi_{2n}$. *In particular, a maximal permutation cannot exhibit* π_{2n} *for* $n \geq 3$ *on any tandem block.*

PROOF: We note first that

$$\theta_6 = (1\ 4\ 5\ 2\ 6)(3)$$

forces π_{2n}: if F is θ_6-adjusted on $\mathcal{P} = \{1, \ldots, 6\}$, then the proper itineraries

$$\{[1, 2], [5, 6], [1, 2]\},$$
$$\{[2, 3], [4, 5], [2, 3]\}$$

and

$$\{[3, 4], [4, 5], [3, 4]\}$$

define three orbits of period 2 whose union is a representative of π_6 (see figure 21).

But then, for $n = k + 3$, $k \geq 1$, we find θ_{2n} forcing $\pi_{2n} = \pi_{6+2k}$ by surrounding a copy of θ_6 with k 2-cycles: that is,

$$\theta_{6+2k} = (1\ 6+2k)(2\ 6+2k-1)\ldots(k\ 8+k)(k+1\ k+4\ k+5\ k+2\ k+6)(k+3).$$

Finally, if a permutation η exhibits π_{2n}, $n \geq 3$, on a tandem block, then by replacing the action on this block with the action of θ_{2n} we obtain a different permutation of the same degree which forces η. ∎

We are now in a position to characterize any tandem blocks appearing in maximal elements of $\mathfrak{S}_n$.

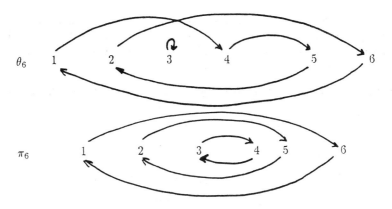

Figure 21

10.12. PROPOSITION. *Suppose θ is forcing-maximal in $\mathfrak{S}_n$. Then:*

(i) *the endpoints of any maximal tandem block for θ are both critical points for θ;*

(ii) *there is at most one maximal tandem block for θ, and the restriction of θ to this block represents either the identity or one of the flips π_2 or π_4 of 10.10.*

PROOF:

Proof of (i): Let $B = \{\alpha, \ldots, \alpha + \beta\}$ be a maximal tandem block for θ, and suppose α is not a critical point for θ. Let $k = per(\alpha)$; by *10.8*, α is not critical for θ^i, $i = 1, \ldots, k$, and it follows that $\alpha > 1$ and that θ^i maps $\{\alpha - 1, \alpha, \alpha + 1\}$ monotonically for $i = 1, \ldots, k$. In particular, since $\theta^k(\alpha) = \alpha < \alpha + 1 = \theta^k(\alpha + 1)$, we must have $\theta^k(\alpha - 1) \leq \alpha - 1$.

Define $\eta \in \mathfrak{S}_n$ by

$$\begin{cases} \eta(i) = \theta(i) & \text{if } i \neq \alpha, \alpha + 1 \\ \eta(\alpha) = \theta(\alpha + 1) \\ \eta(\alpha + 1) = \theta(\alpha) \end{cases}$$

Let F (resp. G) be θ-adjusted (resp. η-adjusted) on $\mathcal{P} = \{1, \ldots, n\}$. Define intervals I_j, J_j, $j = 0, 1$ by

$$I_0 = [\alpha - 1, \alpha], \quad I_1 = [\alpha, \alpha + 1],$$
$$J_0 = \text{clos}\langle\theta(\alpha - 1), \theta(\alpha)\rangle, J_1 = \text{clos}\langle\theta(\alpha), \theta(\alpha + 1)\rangle.$$

Then I_j F-covers J_j with index σ (independent of j); on the other hand I_0 G-covers J_0 and J_1 with index σ, while I_1 G-covers J_1 with index $-\sigma$. Otherwise, every edge in the Markov graph of θ also occurs (with the same index) in the Markov graph of η.

Now, since $\alpha - 1$ is not tandem to α in θ, F^k cannot be monotone on $[\alpha - 1, \alpha]$; let x be the first critical point of F^k to the left of α, so $\alpha - 1 \leq x < \alpha$ and F^k is increasing on $[x, \alpha]$. Since every critical value of F belongs to $\mathcal{P}$, we have $F^k(x) \in \mathcal{P}$, and $F^k(x) < F^k(\alpha) = \alpha$, so $F^k(x) \leq \alpha - 1$. Thus, $[x, \alpha]$ monotonically F^k-covers $[\alpha - 1, \alpha]$, and so for some $y \in [x, \alpha)$ the interval $[y, \alpha]$ is the representative set $\mathcal{R}(F, \mathcal{I}_0)$ of a primitive loop $\mathcal{I}_0$ of length k beginning at I_0. Also,

$$I_1 = \mathcal{R}(F, \mathcal{I}_1)$$

for a loop $\mathcal{I}_1$ of length k starting at I_1. Assume for a moment that $\alpha, \alpha + 1$ do not belong to a self-tandem cycle. Then $\mathcal{I}_1$ is a primitive loop. Our earlier observations on the relation between F-covering and G-covering show that $\mathcal{I}_0$ and $\mathcal{I}_1$ are proper G-itineraries, and it is easily checked that corresponding elements of these two loops are adjacent. Thus, for $j = 0, 1$, $\mathcal{R}(G, \mathcal{I}_j)$ contains a fixedpoint z_j of G^k, and the reader can check that replacing the F-orbits of $\alpha, \alpha + 1$ (as a subset of $\mathcal{P}$) with the G-orbits of z_0, z_1 yields a representative of θ in G, so $\eta \Rightarrow \theta$.

If α, $\alpha + 1$ belong to a self-tandem cycle, then $\beta = 1$, $k = 2\ell$ and the loop $\mathcal{I}_0$ includes the interval $[\alpha + 1, \alpha + 2]$ in the ℓ position. Thus, the fixedpoint $z_0 \in \mathcal{R}(G, \mathcal{I}_0)$ of $G^{2\ell}$ has $G^{\ell}(z_0) \in [\alpha + 1, \alpha + 2]$, and so by replacing $\theta^i(\alpha) \in \mathcal{P}$ with $G^i(z_0)$, $i = 0, \dots, 2\ell$, we again obtain a representative of θ in G, so $\eta \Rightarrow \theta$.

We have showed that if the left endpoint α of a maximal tandem block B for θ is not critical, then θ fails to be forcing-maximal in $\mathfrak{S}_n$. The argument for the right endpoint is similar (in fact, by *10.8* it is even unnecessary if α and $\alpha + \beta$ belong to the same cycle). This proves *(i)*.

Proof of (ii): Note that by *(i)* the endpoints of any maximal tandem block are critical points of opposite type. It follows (since every such block is the image of another such block) that the two endpoints of any maximal tandem block include a maximum value and a minimum value. But then the existence of two disjoint maximal tandem blocks contradicts *10.9*. Thus, the maximal tandem block B is unique.

Since $\theta[B] = B$ and θ is monotone on B, $\theta|B$ is either the identity or a flip; in the latter case, since $\theta = \theta_{**}$, $\theta|B$ represents π_{2j} for some j, and by *10.11*, $j = 1$ or 2, as required. ∎

We can now characterize the forcing-maximal permutations of fixed degree.

10.13. THEOREM. *Suppose $\theta \in \mathfrak{S}_n$. The following conditions are necessary and sufficient for θ to be forced by no other $\eta \in \mathfrak{S}_n$:*

(i) *every maximum value of θ is above every minimum value of θ;*

(ii) *let α be the highest minimum value, and $\alpha + \beta$ the lowest maximum value of θ; then every $i \in \{1, \dots, \alpha, \alpha + \beta, \dots, n\}$ is a critical point of θ;*

(iii) *if $\beta > 1$ in (ii), then $B = \{\alpha, \dots, \alpha + \beta\}$ is the unique maximal tandem block for θ and $\theta|B$ represents the identity or the flip π_4.*

PROOF: Necessity of *(i)-(iii)* follows immediately from *10.8-10.12*.

For sufficiency, suppose $\theta \in \mathfrak{S}_n$ satisfies *(i)-(iii)* and $\mathcal{Q} = \{x_1 < \cdots < x_n\}$ is a representative of θ in F, where F is η-adjusted on $\mathcal{P} = \{1, \dots, n\}$ for some $\eta \in \mathfrak{S}_n$. We need to show $\eta = \theta$.

By *2.1*, we can pick $y_i \in \mathcal{P}$, $i = 1, \dots, \alpha, \alpha + \beta, \dots, n$ such that

(a) $y_1 < x_2$, $y_n > x_{n-1}$, and if

$$i, j \in \{1, \dots, \alpha, \alpha + \beta, \dots, n - 1\}$$

with $i < j$, then
$$\{x_i, y_i\} < \{x_{i+1}, y_{i+1}\};$$

(b) if i is a critical point of σ-maximum type for θ, then y_i is a σ-maximum for F, F is monotone on $\langle x_i, y_i \rangle$, and

$$F(y_i) \geq_\sigma F(x_i).$$

Let $\{z_i \mid i = 1, \cdots \alpha, \alpha + \beta, \ldots, n\}$ be the F-images of the y_i's, numbered so that

$$z_1 < z_2 < \cdots < z_\alpha < z_{\alpha+\beta} < \cdots < z_n.$$

Our proof of the theorem rests on the following

CLAIM : $z_i = i = y_i$ for $i = 1, \ldots, \alpha, \alpha+\beta, \ldots, n$, and $x_i = i$ for $i = \alpha, \ldots, \alpha+\beta$ if $\beta > 1$.

Case 1: $\beta = 1$: In this case, $z_1 < \cdots < z_n$ (and $y_1 < \cdots < y_n$) are n distinct points of $\mathcal{P} = \{1, \ldots, n\}$, so $z_i = i = y_i$ by pigeonholing, and there is no claim about x_i.

Case 2: $\beta = 3$ and $\theta | B$ is π_4.: Here, we have $\theta(\alpha+j) = \alpha+3-j$ for $j = 0, 1, 2, 3$. Suppose $x_{\alpha+1}$ (and hence $x_{\alpha+2} = F(x_{\alpha+1})$) does not belong to $\mathcal{P}$. Since distinct points of $\mathcal{Q} \setminus \mathcal{P}$ have distinct itineraries, $\mathcal{P}$ must intersect $(x_{\alpha+1}, x_{\alpha+2})$ and at least one of $(x_\alpha, x_{\alpha+1})$, $(x_{\alpha+2}, x_{\alpha+3})$. Denote these two points of intersection by $z_{\alpha+1} < z_{\alpha+2}$. Then we have defined

$$\{z_1 < z_2 < \cdots < z_n\} = \{1 < \cdots < n\},$$

so $z_i = i$ for all i. We need to show $x_i = z_i$ for $i = \alpha, \alpha+1, \alpha+2, \alpha+3$.

Note that for $i \neq \alpha+1, \alpha+2$, z_i is the image of some element of $\mathcal{P} \setminus \{z_{\alpha+1}, z_{\alpha+2}\}$. Thus, the set $\mathcal{P} \setminus \{z_{\alpha+1}, z_{\alpha+2}\}$ is F-invariant. Since F is $\mathcal{P}$-adjusted, this means $[z_{\alpha+1}, z_{\alpha+2}]$ is F-invariant. However, one of $\{x_{\alpha+1}, x_{\alpha+2}\}$ but not the other (which is its F-image) belongs to this interval–a contradiction, showing $x_i = z_i$ for $i = \alpha+1, \alpha+2$. Note also that $y_{\alpha+1} = z_{\alpha+1} = \alpha+1$, $y_{\alpha+2} = z_{\alpha+2} = \alpha+2$ gives us $y_i = i \in \mathcal{P}$ for $i = 1, \ldots, n$, in particular $y_\alpha = \alpha$ and $y_{\alpha+3} = \alpha+3$. But

$$\alpha = z_\alpha = F(y_{\alpha+3}) \le F(x_{\alpha+3}) = x_\alpha < x_{\alpha+1} = \alpha+1$$

and

$$\alpha+3 = z_{\alpha+3} = F(y_\alpha) \ge F(x_\alpha) = x_{\alpha+3} > x_{\alpha+2} = \alpha+2.$$

Since F is $\mathcal{P}$-adjusted, $(\alpha, \alpha+1) \cup (\alpha+2, \alpha+3)$ contains no periodic orbits of F. Thus, we must have $x_\alpha = z_\alpha$ and $x_{\alpha+3} = z_{\alpha+3}$.

Case 3: Assume now that B consists of fixedpoints. We have, for $\ell = \beta$,

$$z_1 < z_2 < \cdots < z_\alpha \le x_\alpha < \cdots < x_{\alpha+\ell} \le z_{\alpha+\ell} < \cdots < z_n.$$

Suppose we have this for some ℓ, $1 \le \ell \le \beta$; we will show that there exists $z_{\alpha+\ell-1} \in \mathcal{P}$ such that $x_{\alpha+\ell-1} \le z_{\alpha+\ell-1} < z_{\alpha+\ell}$, and $z_{\alpha+\ell} = x_{\alpha+\ell}$. Note that since the x_i's are fixedpoints, unless both x_i and x_{i+1} belong to $\mathcal{P}$, the open interval (x_i, x_{i+1}) intersects $\mathcal{P}$. Thus in any case each of the half-open intervals $(x_i, x_{i+1}]$, $[x_i, x_{i+1})$ contains at least one point of $\mathcal{P}$. Picking one point of $\mathcal{P}$ from each of $(x_{\alpha+i}, x_{\alpha+i+1}]$, $i = 0, \ldots, \ell-1$, we obtain ℓ points of $\mathcal{P}$ which are distinct from the $n+1-\ell$ z_i's unless $z_{\alpha+\ell} = x_{\alpha+\ell}$. Since $\mathcal{P}$ has only n points, this forces $z_{\alpha+\ell} = x_{\alpha+\ell}$, and we can then pick $z_{\alpha+\ell-1} \in [x_{\alpha+\ell-1}, x_{\alpha+\ell})$. Repeating the process, we get down to $\ell = 1$:

$$z_1 < z_2 < \cdots < z_\alpha \le x_\alpha < x_{\alpha+1} \le z_{\alpha+1} < \cdots < z_n,$$

with $z_{\alpha+i} = x_{\alpha+i}$ for $i = 2, \ldots, \beta$. Now, unless $z_\alpha = x_\alpha$ and $z_{\alpha+1} = x_{\alpha+1}$, the interval $(z_\alpha, z_{\alpha+1})$ must intersect $\mathcal{P}$. Thus, we must have $z_\alpha = x_\alpha$ and $z_{\alpha+1} = x_{\alpha+1}$. This completes case *3* of the claim.

To complete the proof of the theorem, we need to show that

$$F(i) = \theta(i)$$

for $i = 1, \ldots, n$. Note that when $\beta \geq 2$, the claim already gives us, for $i = \alpha, \ldots, \alpha + \beta$, that

$$F(i) = F(x_i) = x_{\theta(i)} = \theta(i).$$

Now, consider the minimum values of θ, which we know are $1, \ldots, \alpha$; consider their θ-preimages, $i_1, \ldots, i_\alpha$: thus,

$$\theta(i_j) = j \quad j = 1, \ldots, \alpha.$$

First, for $j = 1$, we have

$$F(i_1) = F(y_{i_1}) \leq F(x_{i_1}) = x_1 < y_2 = z_2 = 2,$$

so

$$F(i_1) = 1.$$

Now, by induction, we have for $j = 2, \ldots, \beta$,

$$j - 1 = z_{j-1} = F(i_{j-1}) < F(i_j) = F(y_{i_j}) \leq F(x_{i_j}) = x_j < y_{j+1} = z_{j+1} = j+1$$

and hence

$$F(i_j) = j \quad \text{for } j = 1, \ldots, \beta.$$

Similarly, for $j = 0, \ldots, n - (\alpha + \beta)$, let

$$\theta(i_{n-j}) = n - j.$$

We have by a similar induction that

$$F(i_{n-j}) = n - j \quad \text{for } j = 1, \ldots, n - (\alpha + \beta).$$

Since we already have

$$F(i) = \theta(i) \in \{\alpha, \ldots, \alpha + \beta\}$$

for $i = \alpha, \ldots, \alpha + \beta$, this gives us

$$\eta(i) = F(i) = \theta(i) \quad \text{for } i = 1, \ldots, n$$

and proves the theorem. ∎

Note that if a permutation θ satisfies *((i)-(iii))* of Theorem *(10.13)*, and $B = \{\alpha, \ldots, \alpha + \beta\}$, then every point to the left (resp. right) of B is a minimum (resp. maximum) value of θ. Since successive critical points alternate type, the number of points on one side of B equals or exceeds by 1 the number on the other side. If θ is (k, σ)-fold, then $k = n - \beta$. Note that k odd implies equal numbers of points on either side of B; if k is even, there is one more σ-minimum than σ-maximum, and the number of points σ-left of B exceeds the number to its σ-right by 1.

For completeness, we present the characterization of maximal elements in $\mathfrak{C}_n$, which is due to Jungreis [**J**]. Lemma *10.7* specializes to cycles, telling us that any cycle which has a non-critical point is forced by another cycle of the same length. Proposition *10.9*, on the other hand, does not guarantee that a cycle with some minimum value exceeding some maximum value is forced by another cycle of the same length. In fact, it is clear that the permutation obtained by composing a cycle of length 3 or more with an interchange of two elements (as in the proof of *10.9*) must have two cycles.

On the other hand, it is possible to compose a cycle with a (cyclic) interchange of three elements to obtain a new cycle. This is described in the following.

10.14. LEMMA. *Suppose $\theta \in \mathfrak{C}_n$, $n \geq 3$, and let x_j $(j = 1, 2, 3)$ be distinct points in $\mathcal{P} = \{1, \ldots, n\}$ numbered so that x_2 occurs between x_1 and x_3 along the cycle: that is,*

$$\theta^{\alpha_j}(x_j) = x_{\xi(j)}$$

where $\xi = (1\ 2\ 3)$, $0 < \alpha_j$ $(j = 1, 2, 3)$, and $\alpha_1 + \alpha_2 + \alpha_3 = n$. Given $\pi \in \mathfrak{S}_3$, define $\eta \in \mathfrak{S}_n$ by

$$\eta(i) = \begin{cases} \theta(x_{\pi(j)}) & \text{if } i = x_j \ (j = 1, 2, 3) \\ \theta(i) & \text{otherwise.} \end{cases}$$

Then η is a cycle iff $\pi = \xi$ or $\pi = id$.

PROOF: Our hypothesis on x_j and α_j means

$$\theta^k(x_j) \notin \{x_1, x_2, x_3\} \quad \text{for } 1 \leq k < \alpha_j.$$

Thus, if $j = 1, 2, 3$ and $1 \leq k \leq \beta_j \underset{def}{=} \alpha_{\pi(j)}$, then

$$\eta^k(x_j) = \eta^{k-1}\theta(x_j) = \theta^k(x_{\pi(j)}).$$

In particular,

$$\eta^{\beta_j}(x_j) = x_{\xi(\pi(j))}, \quad j = 1, 2, 3.$$

Now, η is a cycle iff x_1, x_2 and x_3 belong to the same η-orbit.

If η is a cycle, then $\xi \circ \pi \in \mathfrak{S}_3$ cannot have a fixedpoint, since $\xi \circ \pi(j) = j$ implies that x_j belongs to a subcycle of η with length $\beta_j < n$. Hence, $\xi \circ \pi$ equals either ξ or $\xi^2 = (1\ 3\ 2)$. But then multiplication on the left by $\xi^{-1} = \xi^2$ gives $\pi = id$ or $\pi = \xi$.

Conversely, it is trivial that $\pi = id$ gives $\eta = \theta \in \mathfrak{C}_n$. If $\pi = \xi$, then for $j = 1, 2, 3$

$$\eta^{\beta_j}(x_j) = x_{\xi^2(j)}$$

or, more precisely,

$$\eta^{\alpha_2}(x_1) = \eta^{\beta_1}(x_1) = x_3$$
$$\eta^{\alpha_1}(x_3) = \eta^{\beta_3}(x_3) = x_2$$

so x_1, x_2 and x_3 lie on a single η-orbit, and η is cyclic. ∎

To apply *10.14*, it will be useful to have a notation for the time-order of points in a cycle. Suppose $\theta \in \mathfrak{S}_n$ $(n \geq 3)$, $\mathcal{P} = \{1, \ldots, n\}$, and $\mathcal{Q} \subset \mathcal{P}$. If $x, y \in \mathcal{Q}$, we write

$$x \underset{\mathcal{Q}}{\rightarrow} y$$

if $\theta^i(x) \notin \mathcal{Q}$ for $i = 1, \ldots, k - 1$ and $\theta^k(x) = y$. The hypotheses of *10.14*, with $\mathcal{Q} = \{x_1, x_2, x_3\}$, are $x_j \underset{\mathcal{Q}}{\rightarrow} x_{\xi(j)}$, $j = 1, 2, 3$. Using *10.14* (and this notation), we obtain the following.

10.15. LEMMA. *Suppose $\theta \in \mathfrak{C}_n$ $(n \geq 3)$ and $\mathcal{Q} = \{x_1, x_2, x_3\} \subset \mathcal{P} = \{1, \ldots, n\}$ is a set of distinct critical points satisfying, for $\sigma = \pm 1$,*

(a) $\theta(x_1) <_\sigma \theta(x_2) <_\sigma \theta(x_3)$
(b) $\theta(x_j) \geq_\sigma \theta[\mathcal{N}(x_j)]$ *for $j = 1, 2$*
(c) $\theta(x_3) \leq_\sigma \theta[\mathcal{N}(x_3)]$.
(d) $x_j \underset{\mathcal{Q}}{\rightarrow} x_{\xi(j)}$ *for $j = 1, 2, 3$, where $\xi = (1\ 2\ 3)$.*

Then there exists $\eta \in \mathfrak{C}_n$ which forces but does not equal θ.

PROOF: We define $\eta \in \mathfrak{S}_n$ as in *10.14*, by

$$\eta(i) = \begin{cases} \theta(x_{\xi(j)}) & \text{if } i = x_j, \ j = 1, 2, 3 \\ \theta(i) & \text{otherwise.} \end{cases}$$

By *(d)* and *10.14*, $\eta \in \mathfrak{C}_n$, and clearly $\eta \neq \theta$. We need to show $\eta \Rightarrow \theta$. This is very similar to the proof of *10.9*. Note that $\eta(x_j) >_\sigma \theta(x_j)$ for $j = 1, 2$, while $\eta(x_3) <_\sigma \theta(x_3)$. Let F be η-adjusted, and G θ-adjusted, on $\mathcal{P} = \{1, \ldots, n\}$. Then we see immediately that if I, J are $\mathcal{P}$-intervals and I G-covers J, then also I F-covers J, so every proper loop for G is also a proper loop for F with the same orientations. Let $u = \theta(x_2)$, and define the itineraries $\mathcal{I}_\pm = \{I_0^\pm, \ldots, I_n^\pm\}$ for G by: $I_0^\pm = \text{clos}\langle u, u - \sigma \rangle$ and $I_{j+1}^\pm$ is the unique $\mathcal{P}$-interval containing $\theta^{j+1}(u)$ and G-covered by $I_j^\pm$. Note that by *(a)*, u is not an endpoint, so both itineraries are defined, while by *(b)* we have $I_n^+ = I_n^- = I_0^-$. Thus, $\mathcal{I}_-$ is a loop for G (a translation of the fundamental loop of θ) of length $k = n$ or $k = \frac{n}{2}$, the latter iff θ is a doubling. Note that I_{n-1}^+ F-covers I_0^+, so that $\mathcal{I}_+$ is a proper loop for F of length n and is prime. One can check that $\mathcal{R}(\mathcal{I}_+, F)$ contains a representative of θ or of a doubling of θ; in either case, F exhibits θ, so $\eta \Rightarrow \theta$. ∎

A corollary of *10.15* is the following set of necessary conditions for $\theta \in \mathfrak{C}_n$ $(n \geq 3)$ to be maximal in $\mathfrak{C}_n$.

10.16. PROPOSITION. *Suppose θ is maximal in $\mathfrak{C}_n$. Let the maximum values of θ be*

$$\{\theta(u_1) < \theta(u_2) < \cdots < \theta(u_k)\}$$

while the minimum values are

$$\theta(v_1) > \theta(v_2) > \cdots > \theta(v_\ell).$$

Then

(i) $\theta(u_2) > \theta(v_2)$.
(ii) *If $\theta(u_1) < \theta(v_i)$, $i \geq 2$, then*

$$v_i \underset{\mathcal{Q}}{\to} v_{i-1} \underset{\mathcal{Q}}{\to} \cdots \underset{\mathcal{Q}}{\to} v_1 \underset{\mathcal{Q}}{\to} u_1 \underset{\mathcal{Q}}{\to} v_i$$

where $\mathcal{Q} = \{u_1, v_1, \ldots, v_i\}$.
(iii) *If $\theta(v_1) > \theta(u_i)$, $i \geq 2$, then*

$$u_i \underset{\widetilde{\mathcal{Q}}}{\to} u_{i-1} \underset{\widetilde{\mathcal{Q}}}{\to} \cdots \underset{\widetilde{\mathcal{Q}}}{\to} u_1 \underset{\widetilde{\mathcal{Q}}}{\to} v_1 \underset{\widetilde{\mathcal{Q}}}{\to} u_i$$

where $\widetilde{\mathcal{Q}} = \{u_1, \ldots, u_i, v_1\}$.

PROOF:

Proof of (i): If *(i)* fails, we have

$$\theta(u_1) < \theta(u_2) < \theta(v_2) < \theta(v_1).$$

Define $a, b, c \in \{1, \ldots, n\}$ by

$$\theta^a(u_1) = u_2$$
$$\theta^b(u_1) = v_2$$
$$\theta^c(u_1) = v_1.$$

If $a < \max(b,c)$, we can apply 10.15 with $x_1 = u_1$, $x_2 = u_2$ and $x_3 = \theta^{\max(b,c)}(u_1)$ (and $\sigma = +1$) to find $\eta \Rightarrow \theta$, $\theta \neq \eta \in \mathfrak{C}_n$. Thus, we can assume $a > \max(b,c)$. Iff $a > b > c$, then similarly 10.15 with $x_1 = v_1$, $x_2 = v_2$, and $x_3 = u_1$ or u_2 gives $\theta \neq \eta \in \mathfrak{C}_n$, $\eta \Rightarrow \theta$. Thus, we can assume $a > c > b$. That is, if we let $Q = \{u-1, u_2, v_1, v_2\}$, then for θ we have $u_1 \underset{Q}{\to} v_2 \underset{Q}{\to} v_1 \underset{Q}{\to} u_2 \underset{Q}{\to} u_1$.

Now, define $\eta \in \mathfrak{S}_n$ by

$$\eta(i) = \begin{cases} \theta(v_j) & \text{if } i = u_j, \ j = 1,2 \\ \theta(u_j) & \text{if } i = v_j, \ j = 1,2 \\ \theta(i) & \text{otherwise.} \end{cases}$$

Then one can check that for η we have

$$u_1 \underset{Q}{\to} u_2 \underset{Q}{\to} v_1 \underset{Q}{\to} v_2 \underset{Q}{\to} u-1$$

so that $\theta \neq \eta \in \mathfrak{C}_n$, while $\eta(u_j) > \theta(u_j)$ and $\eta(v_j) < \theta(v_j)$ for $j = 1,2$, so that an itinerary argument as in 10.15 gives $\eta \Rightarrow \theta$.

Proof of (ii) and (iii): If $\theta(u_1) < \theta(v_i)$ and $u_1 \underset{\widetilde{Q}}{\to} v_a \underset{\widetilde{Q}}{\to} v_b$ where $1 \leq a < b \leq i$, then $\theta(u_1) < \theta(v_b) < \theta(v_a)$, and 10.15 applies with $x_1 = v_a$, $x_2 = v_b$, $x_3 = u_1$ and $\sigma = -1$, to show η is not maximal.

Similarly, if $\theta(v_1) > \theta(u_i)$ and $v_1 \underset{Q}{\to} u_a \underset{Q}{\to} u_b$, $1 \leq a < b \leq i$, then 10.15 applies with $\sigma = +1$, $x_1 = v_1$, $x_2 = u_a$, $x_3 = u_b$. ∎

Now we can characterize forcing-maximal cycles, following Jungreis:

10.17. Theorem. **[J]** $\theta \in \mathfrak{C}_n$ *is not forced by any other cycle of the same degree if and only if*

(i) θ *is* $(n-1)$*-fold (i.e., every point is critical), and*

(ii) *let* $\theta(u_1)$ *be the lowest maximum value, and* $\theta(v_1)$ *the highest minimum value, of* θ; *if the values of* θ *lying in* $[\theta(u_1), \theta(v_1)]$ *are* $\theta(u_i)$, $i = 1, \ldots, k$ (*maxima*) *and* $\theta(v_i)$, $i = 1, \ldots, \ell$ (*minima*), *numbered so that* $\theta(u_i) < \theta(u_{i+1})$ *and* $\theta(v_i) > \theta(v_{i+1})$, *then*

$$\theta(u_1) < \theta(v_\ell) < \cdots < \theta(v_2) < \theta(u_2) < \cdots < \theta(u_k) < \theta(v_1)$$

and, letting $Q = \{u_i \mid i = 1, \ldots, k\} \cup \{v_i \mid i = 1, \ldots, \ell\}$, *we have*

$$u_k \underset{Q}{\to} u_{k-1} \underset{Q}{\to} \cdots \underset{Q}{\to} u_2 \underset{Q}{\to} u_1 \underset{Q}{\to} v_\ell \underset{Q}{\to} v_{\ell-1} \underset{Q}{\to} \cdots \underset{Q}{\to} v_1.$$

Note that in condition *(ii)*, $[\theta(u_1), \theta(v_1)]$ may be empty (if all maximum values exceed all minimum values), in which case *(ii)* is vacuously satisfied.

Proof: The necessity of *(i)* and *(ii)* is essentially already established: necessity of *(i)* is a special case of 10.7, while necessity of *(ii)* follows immediately from 10.16.

Thus, we need to show sufficiency: that every cycle $\theta \in \mathfrak{C}_n$ satisfying *(i)* and *(ii)* is maximal in $\mathfrak{C}_n$. We have already noted that a cycle satisfying *(i)* cannot be self-tandem (i.e., θ cannot be a doubling); thus when *(ii)* is vacuous then by 10.13 θ is maximal in $\mathfrak{S}_n$, hence it is automatically maximal in $\mathfrak{C}_n$. Our proof of the general case (when *(ii)* is not vacuous) is a modification of the proof of 10.13.

COMBINATORIAL PATTERNS FOR MAPS OF THE INTERVAL

Suppose, then, that $\theta \in \mathfrak{C}_n$ satisfies *(i)* and *(ii)* and is forced by another cycle η of the same degree. We can assume η is maximal in $\mathfrak{C}_n$, so it also satisfies *(i)* and *(ii)*. Using *2.1*, we see that θ and η are both $(n-1, \sigma)$-fold for the same $\sigma \in \{\pm 1\}$.

Extend the notation of *(ii)* for θ, letting the maximal values of θ be $\theta(u_1) < \theta(u_2) < \cdots < \theta(u_\alpha)$ and letting the minimum values of θ be $\theta(v_1) > \theta(v_2) > \cdots > \theta(v_\beta)$; thus we have $\alpha \geq k$ and $\beta \leq \ell$.

Our assumption *(ii)* on θ gives us the space order

(a)
$$\begin{aligned}
\theta(v_\beta < \theta(v_{\beta-1}) < \cdots < \theta(v_{\ell+1} < \\
< \theta(u_1) < \theta(v_\ell) < \cdots < \theta(v_2) < \\
< \theta(u_2) < \cdots < \theta(u_k) < \theta(v_1) < \\
< \theta(u_{k+1}) < \cdots < \theta(u_\alpha)
\end{aligned}$$

and

(b)
$$\begin{aligned}
v_1 \xrightarrow[\varrho]{} u_k \xrightarrow[\varrho]{} u_{k_1} \xrightarrow[\varrho]{} \cdots \xrightarrow[\varrho]{} u_1 \xrightarrow[\varrho]{} v_\ell \\
\xrightarrow[\varrho]{} v_{\ell-1} \xrightarrow[\varrho]{} \cdots \xrightarrow[\varrho]{} v_1 \xrightarrow[\varrho]{} u_k.
\end{aligned}$$

Note that u_i (resp. v_i) are also maxima (resp. minima) of η.

Let F be η-adjusted on $\mathcal{P} = \{1, \ldots, n\}$ and let $\mathcal{R} = \{r(1) < \cdots < r(n)\}$ be a representative of θ in F. By *2.1(i)*, for each $i \in \{1, \ldots, n-2\}$ the open interval $(r(i), r(i+2))$ must contain a turning point for F of the same type as $r(i+1)$ (for θ). Since $1 \leq f(1)$ and $r(n) \leq n$, we have

$$r(i) \leq i + 1 \leq r(i+2) \quad \text{for } i = 1, \ldots, n-2.$$

Now, we know from *(a)* that $\theta(u_\alpha) = n$. Thus, $F[r(u_\alpha)] = r(n)$; since $u_\alpha \in \mathcal{P}$ is a maximum of F with $F(u_\alpha) \geq F(r(u_\alpha))$, we must have $F(u_\alpha) = n$, so $\eta(u_\alpha) = n$. Inductively, we have

$$\eta(u_j) = \theta(u_j) \quad \text{for } j = k+1, \ldots, \alpha$$

and similarly we get

$$\eta(v_j) = \theta(v_j) \quad \text{for } j = \ell+1, \ldots, \beta.$$

That is, we have

(c)
$$\eta = \theta \text{ on } \mathcal{P} \setminus \mathcal{Q}.$$

We can push part of the argument further: for each j, u_j is a maximum for θ, hence for η, hence for F, and

$$F(u_j) \geq F(r(u_j)) \quad j = 1, \ldots, \alpha.$$

Similarly, v_j is a minimum for θ, η, and F, and

$$F(v_j) \leq F(r(v_j)) \quad j = 1, \ldots, \beta.$$

In particular, it follows that for all relevant j,

(d)
$$\begin{cases} \eta(u_j) \geq \theta(u_j) \\ \eta(v_j) \leq \theta(v_j). \end{cases}$$

It follows from *(a)*, *(c)* and *(d)* that

$$\theta(u_j) \leq \eta(u_j) \leq \theta(u_j) + 1 \quad \text{for } j = 2, \ldots, k$$
$$\theta(v_j) \geq \eta(v_j) \geq \theta(v_j) - 1 \quad \text{for } j = 2, \ldots, \ell.$$

Suppose $\eta(u_j) \neq \theta(u_j)$ for some j; we can assume equality for all higher j. Then we have $\eta(u_j) = \theta(u_j) + 1$, and referring to *(b)*, by *(c)* we get that

$$u_j \underset{\varrho}{\rightarrow} u_j \quad \text{for } \eta,$$

contradicting the cyclicity of η. Thus, we have $\eta(u_j) = \theta(u_j)$ for $j = 2, \ldots, \alpha$. Similarly, we get

$$\eta(v_j) = \theta(v_j) \text{ for } j = 2, \ldots, \beta.$$

This leaves only the possibility that $\eta(v_1) = \theta(u_1)$ and $\eta(u_1) = \theta(v_1)$; but by inspection this gives η more than one cycle, as well.

Thus we must have $\eta = \theta$, showing θ is maximal in $\mathfrak{C}_n$. ∎

11. Entropy Estimates

In this section, we consider entropy estimates associated to patterns. It is well known [**BGMY**] that the topological entropy of any map exhibiting θ is bounded below by the entropy of any θ-monotone map; this entropy in turn equals [**MS**] the logarithm of the spectral radius of the transition matrix of the Markov graph of θ.

11.1. DEFINITION. *The* **entropy of a pattern** $\theta \in \mathfrak{P}$ *is*

$$h(\theta) = \inf\{h(f) \mid f \text{ exhibits } \theta\}.$$

We have noted that the following all equal $h(\theta)$:

$$h(\theta) = \ln[spec(A)] = h(F),$$

where A is the transition matrix for $\mathfrak{M}(\theta)$, $spec(\ )$ denotes spectral radius, and F is any θ-monotone map. Another calculation of $h(\theta)$ is a kind of dual to the definition.

Recall the definition of the k-horseshoe patterns $\mathfrak{h}(k, \sigma) \in \mathfrak{P}_{k+1}$ from §5. It is easy to see that

$$h(\mathfrak{h}(k, \sigma)) = \ln k.$$

Now, given $f \in \mathcal{E}(I)$ and $n \in \mathbb{N}$, let

$$\mathcal{K}_n(f) = \sup\{k \mid \mathfrak{h}(k, \sigma) \in \mathfrak{P}(f^n) \text{ for some } \sigma = \pm 1\}.$$

By *5.2*, our definition of k-horseshoe is equivalent to that in [**MS**], where it is shown that for any $f \in \mathcal{E}(I)$,

$$h(f) = \limsup_{n \to \infty} \frac{1}{n} \ln \mathcal{K}_n(f).$$

Using this, we can establish the " dual" calculation of entropy. The case $\mathcal{R} = \mathfrak{C}$ of the following theorem was found independently by Block-Coven [**BlCv2**] as well as by Takahashi[**T**].

11.2. THEOREM.

(i) *For* $f \in \mathcal{E}(I)$, *let* $\mathcal{R}_n$ *denote any one of* $\mathfrak{C}_n$, $\mathfrak{S}_n$, *or* $\mathfrak{P}_n$ *and* $\mathcal{R}_n(f) = \{\theta \in \mathcal{R}_n \mid f \text{ exhibits } \theta\}$. *Then*

$$h(f) = \limsup_{n \to \infty}\{h(\theta) \mid \theta \in \mathcal{R}_n(f)\}.$$

(ii) *For* $\theta \in \mathfrak{P}$,

$$h(\theta) = \sup\{h(\eta) \mid \eta \in \mathfrak{C}^*(\theta)\} = \sup\{h(\eta) \mid \eta \in \mathfrak{S}^*(\theta)\}.$$

PROOF:

Proof of (i): Note first that for any f,

$$h(f) \geq \max\{h(\theta) \mid \theta \in \mathfrak{P}_n(f)\}$$
$$\geq \max\{h(\theta) \mid \theta \in \mathfrak{S}_n(f)\} \geq \max\{h(\theta) \mid \theta \in \mathfrak{C}_n(f)\}.$$

Thus it will suffice to prove

$$h(f) \leq \limsup_{n \to \infty}\{h(\theta) \mid \theta \in \mathfrak{C}_n(f)\}.$$

Now suppose $\mathcal{K}_n(f) \geq k$. By *5.6*, there exists $\eta \in \mathfrak{C}_{4(k-2)}(f^n)$ such that any map g exhibiting η has $\mathcal{K}_n(g) \geq k-2$. Note that if x is periodic under f^n with least period m then x is periodic for f with least period $\ell \geq m$, and ℓ divides mn. Thus, if $\eta \in \mathfrak{C}_m(f^n)$, there exists $\theta \in \mathfrak{C}_\ell(f)$ such that $\eta \subseteq \theta^n$. Let F be θ-monotone; then F^n exhibits η, hence $\mathcal{K}_n(F) \geq k-2$. Thus $h(\theta) \geq \frac{1}{n}\ln(k-2)$, and $\theta \in \mathfrak{C}_\ell(f)$, $\ell \geq m$. From this it follows that

$$\limsup_{n \to \infty}\{h(\theta) \mid \theta \in \mathfrak{C}_n(f)\} \geq \limsup_{n \to \infty} \frac{1}{n}\ln[\mathcal{K}_n(f) - 2].$$

But the last lim sup equals $\limsup_{n \to \infty} \frac{1}{n}\mathcal{K}_n(f) = h(f)$. Thus, we have the required inequality, proving *(i)*.

Proof of (ii): This is an easy consequence of *(i)*. Let F be θ-adjusted. Then we have, using *(i)*, that

$$h(\theta) = \limsup_{n \to \infty}\{h(\eta) \mid \eta \in \mathfrak{C}_n(F)\} = \limsup_{n \to \infty}\{h(\eta) \mid \eta \in \mathfrak{S}_n(F)\}.$$

But note that $\mathfrak{C}_n(F) = \mathfrak{C}_n(\theta)$ and $\mathfrak{S}_n(F) = \mathfrak{S}_n(\theta)$; furthermore, if $\theta \in \mathfrak{P}_N$, then for $n > N$, $\theta \notin \mathfrak{C}_n(\theta)$. Thus given $\varepsilon > 0$, there exists $\eta \in \mathfrak{C}_n(\theta)$, n arbitrarily large, with $h(\eta) > h(\theta) - \varepsilon$. But $\theta \Rightarrow \eta$ and $n > N$ implies $\theta \Rrightarrow \eta$. Thus we get *(ii)*. ∎

Next, we relate $h(\theta)$ to $h(\eta)$ when $\theta, \eta \in \mathfrak{P}$ and $\theta \Rightarrow \eta$. Clearly $h(\theta) \geq h(\eta)$. We wish to see when the inequality is strict.

11.3. THEOREM. *Suppose $\theta \in \mathfrak{P}$ is irreducible and $\theta \notin \mathfrak{C}_2$. If θ strongly forces $\eta \in \mathfrak{P}$, then $h(\theta) > h(\eta)$.*

Before proving the theorem, we establish the following observations:

11.4. LEMMA. *If $\theta \in \mathfrak{P}_n$ is irreducible, then*

(i) *$\theta(i) \neq \theta(i+1)$ for all $i = 1, \ldots, n-1$, unless θ is one of the two patterns in $\mathfrak{P}_2$ given by a fixedpoint with a single other preimage.*

(ii) *θ has no tandem cycles unless θ is either one of the two elements of $\mathfrak{S}_2$ or the element $(1\ 3)(2)$ of $\mathfrak{S}_3$.*

(iii) *If θ is irreducible and θ strongly forces η, then $\theta(i) \neq \theta(i+1)$ for $i = 1, \ldots, n-1$, and θ has no tandem cycles unless $\theta \in \mathfrak{C}_2$ and $\eta \in \mathfrak{C}_1$.*

PROOF:

Proof of (i): Suppose $\theta \in \mathfrak{P}_n$ has $\theta(i) = \theta(i+1)$ for some $i \in \{1, \ldots, n\}$. Then at most one of $i, i+1$ belongs to the cycle. Let $\mathcal{B}$ be the block structure consisting of single points except for the pair $i, i+1$. The reduction π of θ induced by $\mathcal{B}$ belongs to $\mathfrak{P}_{n-1}$; thus it is nontrivial unless $n = 2$. If $n = 2$, $\theta(1) = \theta(2) = 1$ or 2, and we have the exception listed in *(i)*; otherwise, θ was reducible.

Proof of (ii): Suppose θ has tandem cycles, say $\theta^j(i+1) = \theta^j(i)\pm1$ and $\theta^k(i) = i$, $\theta^k(i+1) = i+1$. If $k = per(i) = per(i+1)$, form $\mathcal{B}$ by taking $\{\theta^j(i), \theta^j(i+1)\}$, $j = 0,\ldots,k-1$ as blocks and otherwise taking single points. If $per(i) \neq per(i+1)$, then assuming i is the leftmost point among the two cycles, we must have $per(i) = 2per(i+1) = 2\ell$, $\theta^j(i+1)$ lies between $\theta^j(i)$ and $\theta^j(i+2)$ for all j, and all of these points for $j = 0,\ldots,\ell-1$ are distinct. We then form $\mathcal{B}$ by letting each triple $\{\theta^j(i),\ \theta^j(i+1),\ \theta^j(i+2),\}$ be a block and otherwise taking single points as blocks. Now, if the induced reduction is trivial, then we have in the first case $\theta = (1\ 2)$ or $\theta = (1)(2)$ and in the second case $\theta = (1\ 3)(2)$. These are precisely the exceptions indicated in *(ii)*.

Proof of (iii): We simply note that the exceptional patterns listed in *(i)* and *(ii)* do not strongly force any pattern, except that the unique 2-cycle forces a fixedpoint. ∎

PROOF OF *11.3*: Let F be the canonical θ-adjusted map. By *11.4(iii)*, θ has no flat or tandem blocks, so F is θ-linear with slope a nonzero integer at each point outside $\mathcal{P} = \{1,\ldots,n\}$. Note that irreducible patterns are automatically strongly transitive. Thus by *8.5* given $\varepsilon > 0$, the representative set of any proper itinerary of length at least $N(\varepsilon)$ is an interval of length $\leq \varepsilon$. This means no interval J with nonempty interior can have $F^m[J]$ contained in a single $\mathcal{P}$-interval for all m, so that $F^m[J]$ must contain at least one point of $\mathcal{P}$ in its interior, for m sufficiently large. Applying this same argument to one of the subintervals of $F^m[J]$ with this point of $\mathcal{P}$ as one endpoint, we find that (for larger m) $F^m[J]$ must contain two distinct points of $\mathcal{P}$; hence J must F^m-cover some $\mathcal{P}$-interval for sufficiently large m.

Now, let $\mathcal{Q}$ be a representative of η in F disjoint from $\mathcal{P}$, and denote by $\theta \vee \eta$ the pattern exhibited by F on $\mathcal{P} \cup \mathcal{Q}$. We claim $\theta \vee \eta$ is irreducible. This follows from the preceding paragraph, for if $\mathcal{B}$ is a non-trivial block structure for $\theta \vee \eta$, some $\mathcal{B}$-interval has interior, and hence some image of it is a subset of another $\mathcal{B}$-interval that contains a $\mathcal{P}$-interval. It follows that at least one pair of points in $\mathcal{P}$ belongs to a single block of $\mathcal{B}$. But then the block structure restricted to $\mathcal{P}$ induces a reduction $\pi \neq \theta$ of θ. Since θ is irreducible, π must be a single point. But this means all of $\mathcal{P}$ is contained in one block. Since $\mathcal{Q}$ is in the convex hull of $\mathcal{P}$, this means $\mathcal{B}$ is a trivial block structure for $\theta \vee \eta$, contrary to hypothesis. Thus, $\theta \vee \eta$ is irreducible.

Now, let $\mathfrak{M} = \mathfrak{M}(\theta \vee \eta)$ be the Markov graph of $\theta \vee \eta$; we know $\mathfrak{M}$ is strongly connected. Form a subgraph $\mathfrak{M}_1$ of $\mathfrak{M}$ by including an edge $J_i \rightarrow J_j$ of $\mathfrak{M}$ in $\mathfrak{M}_1$ if and only if for the corresponding $\mathcal{P} \cup \mathcal{Q}$-intervals (also denoted J_i, J_j), the $\mathcal{Q}$-intervals uniquely defined by $I_i \supset J_i$ and $I_j \supset J_j$ are joined by an edge in $\mathfrak{M}_\eta$.

Let us use $h(\mathfrak{M})$ to denote the logarithm of the spectral radius of the transition matrix $A(\mathfrak{M})$ for $\mathfrak{M}$, with the analogous notation for $\mathfrak{M}_1$ and $\mathfrak{M}_\eta$. We claim $h(\mathfrak{M}_1) \geq h(\eta)$. One way to see this is to consider the one-sided shifts Σ_1, Σ_η determined by $\mathfrak{M}_1$ and $\mathfrak{M}_\eta$, respectively. Note that if a vertex of $\mathfrak{M}_1$ has no edges emanating from it or no edges hitting it, then it does not appear in any sequence in Σ_1. Thus, symbols appearing in Σ_1 correspond to $\mathcal{P} \cup \mathcal{Q}$-intervals which are contained in a $\mathcal{Q}$-interval and cover some other such interval. Map these symbols to the symbols appearing in $\mathfrak{M}_\eta$ by inclusion. This induces a 1-block coding, hence a semiconjugacy φ of Σ_1 into Σ_η. We need to show φ is onto. Suppose $I_i \rightarrow I_j$ in $\mathfrak{M}_\eta$; by the natural identification I_i is a $\mathcal{Q}$-interval that F-covers the $\mathcal{Q}$-interval I_j. The union of the $\mathcal{P} \cup \mathcal{Q}$-intervals in I_i covers the union of $\mathcal{P} \cup \mathcal{Q}$-intervals in I_j; in particular, every $\mathcal{P} \cup \mathcal{Q}$-interval $J_j \subset I_j$ is F-covered by at least one $\mathcal{P} \cup \mathcal{Q}$-interval $J_i \subset I_i$. This says: given an edge $I_i \rightarrow I_j$ in $\mathfrak{M}_\eta$, and any J_j with $\varphi(J_j) = I_j$, there exists at least one J_i with

$\varphi(J_i) = I_i$ and $J_i \to J_j$ in $\mathfrak{M}_1$. Inductively, given any finite path $I_1 \to \cdots \to I_k$ in $\mathfrak{M}_\eta$, and any J_k with $\varphi(J_k) = I_k$, there exists a path $J_1 \to \cdots \to J_k$ in $\mathfrak{M}_1$ with $\varphi[J_1 \to \cdots \to J_k] = [I_1 \to \cdots \to I_k]$. In other words, every cylinder set of length k for Σ_η intersects the φ-image of Σ_1. This proves $\varphi(\Sigma_1)$ is dense in Σ_η, so φ is onto. But then the entropy of Σ_1 is larger than or equal to the entropy of Σ_η. Finally, it is well known that these entropies are given by the logarithms of the spectral radii of the corresponding transition matrices. Thus $h(\mathfrak{M}_1) \geq h(\mathfrak{M}(\eta))$, as claimed.

Now, since $\theta \Rightarrow \theta \vee \eta \Rightarrow \theta$, we have $h(\theta) = h(\theta \vee \eta) = h(\mathfrak{M})$, and $h(\mathfrak{M}_1) \geq h(\eta)$. We need to show $h(\mathfrak{M}) > h(\mathfrak{M}_1)$. Note that $A(\mathfrak{M})$ is an irreducible matrix, since $\theta \vee \eta$ is irreducible. Also, $\mathfrak{M}_1$ is a subgraph of $\mathfrak{M}$, so $A(\mathfrak{M})$ dominates $A(\mathfrak{M}_1)$. Thus, by Wielandt's version of the Perron-Frobenius theorem [$\mathbf{W;S}$], $spec[A(\mathfrak{M})] > spec[A(\mathfrak{M}_1)]$ unless $A(\mathfrak{M}) = A(\mathfrak{M}_1)$. But since $\mathcal{Q} \cap \mathcal{P} = \emptyset$ and $\mathcal{Q}$ is contained in the convex hull of $\mathcal{P}$, the leftmost and rightmost $\mathcal{P} \cup \mathcal{Q}$-intervals are not contained in $\mathcal{Q}$-intervals, hence the corresponding rows and columns of $A(\mathfrak{M}_1)$ are all zero, while $A(\mathfrak{M})$, being irreducible, has no row or column of zeroes. Thus $spec[A(\mathfrak{M})] > spec[A(\mathfrak{M}_1)]$, and the relations

$$h(\theta) = h(\theta \vee \eta) = h(\mathfrak{M}) > h(\mathfrak{M}_1) \geq h(\mathfrak{M}(\eta)) = h(\eta)$$

give the desired conclusion. ∎

11.5. COROLLARY. *If $\theta, \eta \in \mathfrak{C}$ and $\theta \Rightarrow \eta$, then $h(\theta) > h(\eta)$ unless θ and η have a common nontrivial reduction.*

PROOF: If θ is irreducible, this is *11.3*, since a cycle strongly forces any other cycle that it forces. If θ has a reduction π and $\theta \Rightarrow \eta$, then by *3.7* either η extends π, contrary to our hypothesis, or $\pi \Rightarrow \eta$. Thus, we can replace θ with any nontrivial reduction π of θ, using $h(\theta) \geq h(\pi) \geq h(\eta)$. But θ has an irreducible reduction, and for this reduction, $h(\pi) > h(\eta)$. ∎

Finally, we consider bounds on $h(\theta)$ for θ a cycle of fixed degree. Lower bounds are well known [$\mathbf{BGMY}$], resulting from the calculation of entropy for forcing-minimal (or *primary*) cycles, which have been completely characterized [$\mathbf{ALS;Bl;BlC;C;H2;ALM}$]. We consider the opposite estimate

$$H(\mathfrak{C}_n) = \max\{h(\theta) \mid \theta \in \mathfrak{C}_n\}.$$

Theorem *10.16* gave Jungreis' characterization of forcing-maximal elements of $\mathfrak{C}_n$; clearly $H(\mathfrak{C}_n) = h(\theta)$ for θ one of these cycles. However, by contrast with the situation for forcing-minimal cycles, different forcing-maximal elements of $\mathfrak{C}_n$ will in general have different entropy, and it is difficult to decide which ones achieve $H(\mathfrak{C}_n)$. Recently, Geller and Tolosa [$\mathbf{GT}$] have succeeded in using the methods we will present in this section to resolve this question for $\mathfrak{C}_n$ when n is odd. Here, we shall not attempt to obtain a formula for $H(\mathfrak{C}_n)$ as a function of n. Instead, we will give an indirect calculation of the asymptotic behavior of $H(\mathfrak{C}_n)$ as $n \to \infty$.

The corresponding maximal entropy estimates $H(\mathfrak{S}_n)$ and $H(\mathfrak{P}_n)$ can also be defined. $H(\mathfrak{P}_n)$ is easy to calculate. We know from *10.2* that a pattern $\theta \in \mathfrak{P}_n$ with $h(\theta) > 0$ is forced by an $(n-1)$-horseshoe; thus

$$H(\mathfrak{P}_n) = h(\mathfrak{h}(n-1, \pm 1)) = \ln(n-1).$$

On the other hand, the calculation of $H(\mathfrak{S}_n)$ involves the same difficulties as that of $H(\mathfrak{C}_n)$. It is easy to see that the permutations without tandem blocks characterized in *10.13* include those that maximize entropy in $\mathfrak{S}_n$, but comparing the entropy of these is still difficult. Our analysis will show, however, that the two quantities $H(\mathfrak{C}_n)$ and $H(\mathfrak{S}_n)$ have the same asymptotic behavior.

11.6. THEOREM. $H(\mathfrak{C}_n) \sim \ln \frac{2n}{\pi}$; more precisely,

$$\lim_{n \to \infty} \frac{1}{n} \exp H(\mathfrak{C}_n) = \lim_{n \to \infty} \frac{1}{n} \exp H(\mathfrak{S}_n) = \frac{2}{\pi}.$$

Before proving 11.6, we make the following observation, showing that each of the sequences $H(\mathfrak{C}_n)$ and $H(\mathfrak{S}_n)$ is strictly increasing with n.

11.7. LEMMA. If $\eta \in \mathfrak{S}_n$, there exists $\theta \in \mathfrak{S}_{n+1}$ with the same number of cycles as η, with $\theta \Rightarrow \eta$.

PROOF: Given $\eta \in \mathfrak{S}_n$, let $a \in \mathcal{P} = \{1, \ldots, n\}$ be the unique element with $\eta(a) = n$, and define $\theta \in \mathfrak{S}_{n+1}$ by

$$\begin{aligned} \theta(i) \quad &= \eta(i) \text{ if } i \neq a, n+1 \\ \theta(a) \quad &= n+1 \\ \theta(n+1) &= n. \end{aligned}$$

Note that the subcycle $\eta_0 = (a_0 \ \ldots \ a_k)$ of η containing $a_k = n$ is replaced by the subcycle $\theta_0 = (a_0 \ \ldots \ a_{k-1} \ n+1 \ n)$ of θ, and all other subcycles (if any) are unchanged. If $k = 0$, we have replaced a fixed point n with a 2-cycle disjoint form $1, \ldots, n-1$, and clearly this forces η.

In general, consider the η-adjusted map F on $\mathcal{P} = \{1, \ldots, n\}$ and the θ-adjusted map G on $\mathcal{P}' = \{1, \ldots, n+1\}$. Let $J_i = [i, i+1]$ for $i = 1, \ldots, n$. Note that we can assume $F|J_i = G|J_i$ except for $i = a-1, a$, and $n-1$. Now by construction, if $a < n$, then F has a maximum at a; G does as well, and $G(a)$ is higher than $F(a)$. It follows that if J_i F-covers J_j, then also J_i G-covers J_j, and $\sigma(F, J_i, J_j) = \sigma(G, J_i, J_j)$. Now, let $\mathcal{I} = \{I_0, \ldots, I_k = I_0\}$ be the unique proper F-loop with $a \in \mathcal{R}(F, \mathcal{I})$ (uniqueness follows since I_1 must contain n, hence $I_1 = J_{n-1}$). Then $\mathcal{I}$ is also a proper loop for G, so some $x \in I_0$ satisfies $G^j(x) \in I_j$, $j = 0, \ldots, k$. By 1.8, either a is the only fixedpoint of F^k in $\mathcal{R}(F, \mathcal{I})$ and $I_0, \ldots, I_{k-1}$ are disjoint, so a represents η_0, or η is a doubling of some ℓ-cycle π, $k = 2\ell$, $I_0 \cup I_\ell, I_1 \cup I_{\ell+1}, \ldots, I_{\ell-1} \cup I_{2\ell-1}$ are disjoint, and π is the only cycle of length ℓ represented in these intervals. Now, the G-orbit of x represents either η_0 or π. Thus, either θ forces the subcycle η_0 or it forces a 2-reduction π of η. Note that in the second case, the orbit of x is a negative repellor for G^ℓ, so by 9.8 θ also forces the doubling η_0 of π determined by the slopes of G along the orbit of x. So we have a representative $\mathcal{Q}_0$ of η_0 in G; furthermore, if η contains other cycles, the periodic orbits representing them in F are still the identical periodic orbits of G, so that if we take $\mathcal{P}_0$ to be the union of these orbits, it is clear that F on $\mathcal{P} = \mathcal{P}_0 \cup \eta_0$ and G on $\mathcal{Q} = \mathcal{P}_0 \cup \mathcal{Q}_0$ represent the same permutation. ∎

11.7 shows the sequences $H(\mathfrak{S}_n)$ and $H(\mathfrak{C}_n)$ to be increasing, and by 11.5, $H(\mathfrak{C}_n)$ is strictly increasing. We turn now to the proof of 11.6. It is quite long, so we outline our approach. Our proof will be in four steps:

(1) *Step 1:* We will bound $H(\mathfrak{C}_n)$ and $H(\mathfrak{S}_n)$ by $spec(A_{k_n})$, where $k_n = n$ or $n+1$ and A_n is a model matrix depending on n.

(2) *Step 2:* We will embed A_n as an operator $\hat{A}_n$ on $L^1[0,1]$, and find an operator C which is the limit of $T_n = \frac{1}{2n} \hat{A}_n$ in the operator norm.

(3) *Step 3:* We will see that $spec(C) = \frac{2}{\pi}$, proving that

$$\limsup_{n \to \infty} \frac{1}{n} \exp H(\mathfrak{C}_n) \leq \limsup_{n \to \infty} \frac{1}{n} \exp H(\mathfrak{S}_n) \leq \frac{2}{\pi}.$$

(4) *Step 4:* We will show equality by producing a sequence of cycles of increasing length whose entropy grows at an asymptotic rate of $\frac{2}{\pi}$.

For step *1*, define $A_n = [a_{ij}]$ to be the $2n \times 2n$ matrix of zeroes and ones defined by

$$a_{ij}^{(n)} = 1 \text{ iff } |i - j| \le n - 1 \text{ and } |i + 1 - 2(n+1)| \le n + 1;$$

$$a_{ij}^{(n)} = 0 \text{ otherwise.}$$

The matrix defined by this has a "diamond" pattern of 1's. For example,

$$A_4 = \begin{pmatrix} 0 & 0 & 0 & 1 & 1 & 0 & 0 & 0 \\ 0 & 0 & 1 & 1 & 1 & 1 & 0 & 0 \\ 0 & 1 & 1 & 1 & 1 & 1 & 1 & 0 \\ 1 & 1 & 1 & 1 & 1 & 1 & 1 & 1 \\ 1 & 1 & 1 & 1 & 1 & 1 & 1 & 1 \\ 0 & 1 & 1 & 1 & 1 & 1 & 1 & 0 \\ 0 & 0 & 1 & 1 & 1 & 1 & 0 & 0 \\ 0 & 0 & 0 & 1 & 1 & 0 & 0 & 0 \end{pmatrix}.$$

A_n is the transition matrix of a directed graph, but not of the Markov graph of any combinatorial pattern. However, any element of $\mathfrak{P}_{2n+1}$ has a Markov graph with $2n$ vertices, so its transition matrix can be compared with A_n.

Thus, fix $\theta \in \mathfrak{S}_{2n+1}$ and let $M(\theta)$ denote the *transpose* of the transition matrix of $\mathfrak{M}(\theta)$; that is, $M(\theta) = [m_{ij}]$, where

$$m_{ij} = 1 \text{ iff } J_j \to J_i \text{ is an edge in } \mathfrak{M}(\theta).$$

We know that $h(\theta) = \ln spec(M)$, and by the Perron-Frobenius theorem, for any positive matrix B we have

$$spec(B) = \lim_{k \to \infty} [v^T B^k v]^{\frac{1}{k}},$$

where v is a column of 1's of appropriate size; note that $v^T B^k v$ is the sum of the entries in the column $B^k v$.

11.8. LEMMA. *For $\theta \in \mathfrak{S}_{2n+1}$, $M = M(\theta) = [A(\mathfrak{M}(\theta))]^T$, A_n the matrix defined above, and v^T a $2n$-row of 1's,*

$$v^T M \le v^T A_n.$$

That is, each row of M has no more 1's than the corresponding row of A_n.

PROOF: Let F be any map θ-monotone on $\mathcal{P} = \{1, \ldots, 2n + 1\}$.

Since $\theta \in \mathfrak{S}_{2n+1}$, there is a unique point $j \in \{1, \ldots, 2n+1\}$ with $\theta(j) = 1$. The only $\mathcal{P}$-intervals which can F-cover $J_1 = [1, 2]$ are those with j as an endpoint; there are at most two of these (two if $1 < j < 2n+1$, one if $j = 1$ or $j = 2n+1$). Thus, the inequality holds for the first rows.

Now, consider what happens as we move down a row. If a $\mathcal{P}$-interval J_j F-covers J_{i+1} but not J_i, then $y = j$ or $j + 1$ must be a minimum for F, with $\theta(y) = i + 1$. Since θ is bijective, there is at most one such minimum y, and it is an endpoint to at most two intervals. Thus, the number of $\mathcal{P}$-intervals F-covering J_{j+1} can exceed the number F-covering J_j by at most two (of course, it could go down). Stated differently, the $(i + 1)^{\text{st}}$ row of M can have at most two more 1's than the ith row. But for $i < n$, the $(i + 1)^{\text{st}}$ row of A_n has precisely two more 1's than the ith. This gives an inductive proof of the inequality for rows $1, \ldots, n$. (Note that by the time we reach row n, the inequality must be strict if either 1 or $2n + 1$ is a minimum point for F.)

Now, repeat the argument starting from the other end, replacing minima with maxima, working from row $2n$ down to row $n + 1$. ∎

The next lemma will help us establish bounds on the spectral radius of $M = M(\theta)$, using the Perron-Frobenius theorem.

11.9. LEMMA. *For* $k = 0, 1, \ldots$, *let* $u(k) = A_n^k v$, *where* M, A_n, *and* v *are as above. Then*

(i) *the coordinates* u_i *of* $u(k)$ *satisfy*

$$u_1 = u_{2n} \leq u_2 = u_{2n-1} \leq \cdots \leq u_n = u_{n+1}.$$

(ii) $Mu(k) \leq A_n u(k)$, *where inequalities are taken coordinatewise.*

PROOF: We use induction on k.

For $k = 0$, $u(0) = v$ and *(i)* holds trivially, with equality everywhere, while *(ii)* follows from *11.8*.

Given *(i)* and *(ii)* for k, we need to show them for $k + 1$. Note that

$$u(k + 1) = A_n u(k)$$

so that the ith entry of $u(k+1)$ is the sum of the middle $2i$ entries of $u(k)$. That is, for $i = 1, \ldots, n$,

$$u_i(k + 1) = u_{2n+1-i}(k + 1) = \sum_{j=n+1-i}^{n+i} u_j(k).$$

In particular, for $i = 1, \ldots, n$,

$$u_{i+1}(k + 1) = u_{2n-i}(k + 1) = u_i(k + 1) + 2u_{i+1}(k)$$
$$\geq u_i(k + 1) = u_{2n+1-i}(k).$$

This gives us *(i)* for $k + 1$.

To get *(ii)*, we note that by *11.8* the ith entry of $Mu(k)$ is the sum of at most

$$j = 2 \cdot \min\{i, \ldots, 2n + 1 - i\}$$

entries of $u(k)$, while *(i)* says that the corresponding entry of $A_n u(k)$ is the sum of the j *middle* entries of $u(k)$. But by *(i)* for k, these are the j *largest* entries of $u(k)$. From this, *(ii)* is immediate for $k + 1$, and induction establishes *11.9*. ∎

Using *11.9*, we complete step *1* by proving the following result.

11.10. PROPOSITION. $spec(M) \leq spec(A_n)$, *where* M, A_n *are as above.*

PROOF: If we use the Perron-Frobenius theorem to calculate the two spectral radii, *11.10* would follow from the following inequality

$$v^T M^k v \leq v^T A_n^k v \quad \text{for } k = 0, 1, \ldots,$$

and this in turn would follow from the following claim:

$$M^k v \leq A_n^k v \quad \text{for } k = 0, 1, \ldots.$$

Note that the claim is trivial for $k = 0$. Supposing now that it holds for k, multiply both sides on the left by M. Since M is a positive matrix, we get

$$M^{k+1} v = M(M^k v) \leq M \cdot A_n^k v = M \cdot u(k).$$

But we showed in *11.9(ii)* that

$$M \cdot u(k) \leq A_n u(k) = u(k + 1) = A_n^k v.$$

This completes the induction on k, proving *11.10.* ∎

We turn now to step *2*.

We embed $\mathbb{R}^n$ into $L^2[0,1]$ by the linear map $E_n : \mathbb{R}^n \to L^2[0,1]$ which takes each element e_i of the standard basis $\{e_1, \ldots, e_n\}$ for $\mathbb{R}^n$ into the characteristic function of the interval $\left[\frac{i-1}{n}, \frac{i}{n}\right)$. Thus, for $v^T = (v_1, \ldots, v_n)$, we have

$$[E_n v](t) = v_i \quad \text{iff } t \in \left[\frac{i-1}{n}, \frac{i}{n}\right), i = 1, \ldots, n.$$

Note that

$$\langle E_n e_i, E_n e_j \rangle = \mu\left\{\left[\frac{i-1}{n}, \frac{i}{n}\right) \cap \left[\frac{j-1}{n}, \frac{j}{n}\right)\right\} = \frac{1}{n}\delta_{ij}$$

where δ_{ij} is the Kronecker delta. Thus, E_n is an isometry up to the scaling factor $n^{-\frac{1}{2}}$:

$$\|E_n v\|_{L^2} = \frac{1}{\sqrt{n}}\|v\|_{\mathbb{R}^n}.$$

Now, the embedding $E_n : \mathbb{R}^n \to L^2[0,1]$ induces an embedding of linear operators $L : \mathbb{R}^n \to \mathbb{R}^n$ to linear operators $\hat{L} : L^2[0,1] \to L^2[0,1]$. We claim that if L is multiplication by the matrix $B = [b_{ij}]$, then $\hat{B}$ is an integral operator

$$(\hat{L}u)(t) = \int_0^1 \alpha(t,s)u(s)ds$$

where

$$\alpha(t,s) = nb_{ij} \quad \text{for } t \in \left[\frac{i-1}{n}, \frac{i}{n}\right) \text{ and } s \in \left[\frac{j-1}{n}, \frac{j}{n}\right).$$

To check that this calculation gives

$$\hat{L}(E_n v) = E_n(Lv),$$

we calculate: for $t \in \left[\frac{i-1}{n}, \frac{i}{n}\right)$,

$$[\hat{L}(E_n v)](t) = \int_0^1 \alpha(t,s)(E_n v)(s)ds = \sum_{j=1}^n \int_{\left[\frac{i-1}{n}, \frac{i}{n}\right)} nb_{ij}v_j ds$$

$$= \sum_{j=1}^n nb_{ij}v_j\left(\frac{j}{n} - \frac{j-1}{n}\right) = \sum_{j=1}^n b_{ij}v_j.$$

In particular, note that since $\alpha(t,s)$ is a bounded measurable function, $\hat{L}$ is a bounded linear operator on $L^2[0,1]$. One can also check that this embedding respects composition of operators (equivalently, multiplication of representing matrices). We summarize this embedding and its properties, including a few new ones, in

11.11. LEMMA.

(i) *For each $n = 1, 2, \ldots$, the embedding $E_n : \mathbb{R}^n \to L^2[0,1]$ via*

$$(E_n v)(t) = v_i \quad \text{iff } t \in \left[\frac{i-1}{n}, \frac{i}{n}\right)$$

preserves inner products up to a factor of n:

$$\langle E_n v, E_n w \rangle_{L^2} = n \langle v, w \rangle_{\mathbb{R}^n}.$$

(ii) *The induced embedding of the algebra of operators is as follows: if* $L : \mathbb{R}^n \to \mathbb{R}^n$ *is represented with respect to the standard basis by* $B = [b_{ij}]$, *then the bounded linear operator* $\hat{L} = \hat{B} : L^2[0,1] \to L^2[0,1]$ *is defined by: given* $u \in L^2[0,1]$ *and* $t \in [0,1]$,

$$(\hat{L}u)(t) = \int_0^1 \alpha_B(t,s)u(s)ds,$$

where $\alpha_B(t,s)$ *is the sum over* i,j *of*

$$n \cdot b_{ij} \cdot \left[\text{characteristic function of } \left[\frac{i-1}{n}, \frac{i}{n} \right) \times \left[\frac{j-1}{n}, \frac{j}{n} \right) \right].$$

(iii) $L \to \hat{L}$ *obeys:* $\|\hat{L}\| = \|L\|$, *so* $spec(\hat{L}) = spec(L)$. *If* B *is a symmetric matrix, then* $\hat{B}$ *is self-adjoint, so in this case*

$$spec(B) = \|B\| = \|\hat{B}\| = spec(\hat{B}).$$

PROOF: *(i)* and *(ii)* were taken care of earlier.

To see *(iii)*, note that equality of operator norms follows from the fact that E_n is conformal, and equality of spectral radii follows from the standard formula $spec(L) = \lim_{k \to \infty} \|L^k\|^{\frac{1}{k}}$. Finally, it is easy to check self-adjointness of $\hat{B}$ when $B = B^T$, and the final formula follows from the spectral theorem. ∎

We will apply *11.11* to the $2n \times 2n$ matrices A_n defined before; they induce positive, self-adjoint operators $\hat{A}_n$ on $L^2[0,1]$ with

$$spec(A_n) = spec(\hat{A}_n) = \|\hat{A}_n\|.$$

This embedding allows us to compare A_n for different n.

Now, we define an operator C on $L^2[0,1]$ as the integral operator whose kernel is the characteristic function of the " diamond" $\Diamond \subset [0,1] \times [0,1]$ (see figure 22) defined by

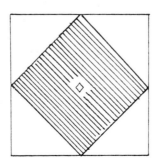

Figure 22: $\Diamond$

$$\Diamond = \{(t, s) \in [0, 1] \times [0, 1] \mid |t + s - 1| \leq \frac{1}{2} \text{ and } |t - s| \leq \frac{1}{2}\}.$$

Thus, for $u \in L^2[0, 1]$ and $t \in [0, 1]$,

$$[Cu](t) = \int_0^1 \alpha_C(t, s) u(s) ds$$

where

$$\alpha_C(t, s) = \begin{cases} 0 & \text{if } |t + s - 1| \leq \frac{1}{2} \text{ and } |t - s| \leq \frac{1}{2} \\ 1 & \text{if } |t + s - 1| > \frac{1}{2} \text{ or } |t - s| > \frac{1}{2}. \end{cases}$$

We now show that C is the operator-norm limit of $T_n = \frac{1}{2n}\hat{A}_n$, with explicit estimates:

11.12. LEMMA. *Let C, $\hat{A}_n$ be as above, and $T_n = \frac{1}{2n}\hat{A}_n$. Then*

$$\|T_n - C\| \leq \frac{1}{\sqrt{2n}}.$$

PROOF: Each operator has as its kernel the characteristic function of a subset of $[0, 1]^2$. The kernel α_C of C is the characteristic function of the diamond $\Diamond$ defined above. The kernel α_n of T_n is, for $t \in \left[\frac{i-1}{2n}, \frac{i}{2n}\right)$, $s \in \left[\frac{i-1}{2n}, \frac{j}{2n}\right)$,

$$\alpha_n(t, s) = \frac{1}{2n} \cdot (2n) a_{ij} = a_{ij} \in \{0, 1\}.$$

Thus, α_n is the characteristic function of the union Δ_n of those squares

$$\left[\frac{i-1}{2n}, \frac{i}{2n}\right) \times \left[\frac{j-1}{2n}, \frac{j}{2n}\right)$$

for which $a_{ij} = 1$ in A_n. From the definition of A_n, we see that $\Diamond \subset \Delta_n$, and $\Delta_n \setminus \Diamond$ consists of triangles obtained from the $4n$ squares in Δ_n that have one or another diagonal on one of the four lines

$$t - s = \pm\frac{1}{2}, \quad t + s - 1 = \pm\frac{1}{2}.$$

Clearly, $T_n - C$ has as its kernel $\alpha_n - \alpha_C$, which is the characteristic function of $\Delta_n \setminus \Diamond$. The Lebesgue measure of this set is

$$\mu(\Delta_n - \Diamond) = \frac{4n}{2}\left[\frac{1}{2n}\right]^2 = \frac{1}{2n}.$$

It follows that

$$\|T_n - C\|_{L^2} \leq [\mu(\Delta_n \setminus \Diamond)]^{\frac{1}{2}} = \frac{1}{\sqrt{2n}}$$

and the lemma is proved. ∎

In fact, as a corollary of the fact that $\Diamond \subset \Delta_n$, we can obtain the more precise estimate

$$\|C\|_{L^2} \leq \|T_n\|_{L^2} \leq \|C\|_{L^2} + \frac{1}{\sqrt{2n}}.$$

Since C and $\hat{A}_n = 2nT_n$ are both self-adjoint, we obtain

11.13. COROLLARY.

$$spec(C) \leq \frac{1}{2n}spec(\hat{A}_n) = \frac{1}{2n}spec(A_n) \leq spec(C) + \frac{1}{\sqrt{2n}}.$$

In particular, $\lim_{n\to\infty} \frac{1}{2n}spec(\hat{A}_n) = spec(C)$.

This completes step 2. Now, we turn to step 3.

11.14. PROPOSITION. $spec(C) = \frac{2}{\pi}$.

PROOF: Let

$$u(t) = \sin(\pi t).$$

We will show that

$$Cu = \frac{2}{\pi}u.$$

Since C is a positive, self-adjoint operator and $u(t)$ is a positive eigenfunction of C, 11.14 will follow by the Perron-Frobenius theorem.

To prove the equality, we calculate

$$[Cu](t) = \int_0^1 \alpha_C(t, s)u(s)ds.$$

For $t < \frac{1}{2}$, this gives

$$[Cu](t) = \int_{\frac{1}{2}-t}^{\frac{1}{2}+t} u(s)ds = \int_{\frac{1}{2}-t}^{\frac{1}{2}+t} \sin(\pi s)ds = -\frac{1}{\pi}\cos(\pi s)\Big|_{\frac{1}{2}-t}^{\frac{1}{2}+t}$$

$$= \frac{1}{\pi}\{\cos[\pi(\frac{1}{2} - t)] - \cos[\pi(\frac{1}{2} + t)]\} = \frac{1}{\pi}\{\cos(\frac{\pi}{2} - \pi t) - \cos(\frac{\pi}{2} + \pi t)\}$$

$$= \frac{1}{\pi}\{\sin(\pi t) - \sin(-\pi t)\} = \frac{2}{\pi}\sin \pi t.$$

Similarly, for $t \geq \frac{1}{2}$,

$$[Cu](t) = \int_{\frac{1}{2}-(1-t)}^{\frac{1}{2}+(1-t)} \sin \pi s ds = \frac{2}{\pi}\sin[\pi(1 - t)] = \frac{2}{\pi}\sin[\pi - \pi t] = \frac{2}{\pi}\sin \pi t. \quad \blacksquare$$

Thus, we have the third step:

11.15. PROPOSITION.

$$\limsup_{n\to\infty} \frac{1}{n}\exp H(\mathfrak{C}_n) \leq \limsup_{n\to\infty} \frac{1}{n}\exp(\mathfrak{S}_n) \leq \frac{2}{\pi}.$$

PROOF: We gather together various results. The first inequality follows since $\mathfrak{C}_n \subset \mathfrak{S}_n$, and $H(\mathfrak{C}_n)$ is a maximum over the former set, while $H(\mathfrak{S}_n)$ is a maximum over the latter. We shall prove the second inequality separately for the subsequence of even and of odd integers.

For any $\theta \in \mathfrak{S}_{2n+1}$, let $M = [A(\mathfrak{M}(\theta)]^T$; then by 11.10

$$\exp h(\theta) = spec(M) \leq spec(A_n).$$

Maximizing over $\theta \in \mathfrak{S}_{2n+1}$, we have

$$\exp H(\mathfrak{S}_{2n+1}) \leq spec(A_n).$$

Then by *11.11* and *11.13*, we have

$$\limsup_{n\to\infty} \frac{1}{2n}\exp H(\mathfrak{S}_{2n+1}) \leq \lim_{n\to\infty}\frac{1}{2n}spec(A_n) = \lim_{n\to\infty}\frac{1}{2n}spec(\hat{A}_n) = spec(C)$$

and by *11.14*

$$\limsup_{n\to\infty} \frac{1}{2n}\exp H(\mathfrak{S}_{2n+1}) \leq \frac{2}{\pi}.$$

Now, by *11.7* we have

$$H(\mathfrak{S}_{2n}) \leq H(\mathfrak{S}_{2n+1})$$

so that on one hand

$$\limsup_{n\to\infty} \frac{1}{2n}\exp H(\mathfrak{S}_{2n}) \leq \frac{2}{\pi}$$

and on the other using $\frac{1}{2n+1} < \frac{1}{2n}$

$$\limsup_{n\to\infty} \frac{1}{2n+1}\exp H(\mathfrak{S}_{2n+1}) \leq \frac{2}{\pi}.$$

This ends step *3*. ∎

Finally, we complete the proof of *11.6* by carrying out step *4*, by finding a sequence of cycles with the right asymptotic growth rate for entropy.

Define $\theta_n \in \mathfrak{P}_{4n+1}$, $n = 1, 2, \dots$ by

$$\theta_n(j) = \begin{cases} 2n+1-j & \text{if } j < 2n \text{ is odd} \\ 2n+j & \text{if } j \leq 2n \text{ is even} \\ j-2n & \text{if } j > 2n \text{ is odd} \\ 6n+3-j & \text{if } j > 2n \text{ is even.} \end{cases}$$

The cases $n = 1$ and $n = 2$ are sketched in figure 23.

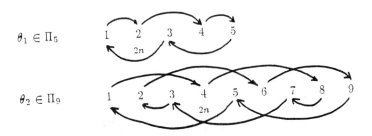

Figure 23

11.16. LEMMA. θ_n *is cyclic.*

PROOF: If $1 \leq j < 2n$ is odd, then

$$\theta_n(j) = 2n+1-j$$

which is even and $\leq 2n$. Thus,

$$\theta_n^2(j) = 2n+\theta_n(j) = 2n+(2n+1-j) = 4n+1-j$$

which is even and $> 2n$. It follows that

$$\theta_n^3(j) = 6n + 3 - \theta_n^2(j) = (6n + 3) - (4n + 1 - j) = 2n + 2 + j,$$

and this is odd and $\geq 2n + 3$; thus

$$\theta_n^4(j) = \theta_n^3(j) - 2n = 2 + j.$$

Thus, θ_n cycles each odd number below $2n$ through the four regimes, landing under θ_n^4 at the next odd number. In particular, starting with $j = 1$, $\theta_n(1)$, $\theta_n^2(1)$, and $\theta_n^3(1)$ are in distinct regimes and then $\theta_n^4(1) = 3$; again, we cycle through getting $\theta_n^8(1) = 5$; and so on with $\theta_n^{4j}(1) = 1 + 2j$, until at $j = n$ we hit the first odd *above* $2n$, namely

$$\theta_n^{4n}(1) = 2n + 1.$$

But then

$$\theta_n^{4n+1}(1) = \theta_n(2n + 1) = 2n + 1 - 2n = 1,$$

and $\theta_n \in \mathfrak{C}_{4n+1}$. ∎

We note that, according to recent (as yet unpublished) work of Geller and Tolosa [**GT**], this cycle θ_n in fact attains the maximum entropy in $\mathfrak{C}_{4n+1}$. However, we need only the following estimate.

11.17. LEMMA.

$$\frac{1}{4n} \exp H(\mathfrak{C}_{4n+1}) \geq \frac{1}{4n} \exp h(\theta_n) \geq \frac{2}{\pi} - \frac{1}{4\sqrt{n}}.$$

PROOF: The first inequality is trivial.

Consider the matrix $[m_{ij}] = M = M(\theta_n) = [A(\mathfrak{M}(\theta_n))]^T$. For j odd, J_j covers J_i whenever

$$i \in \begin{cases} \{2n + 1 - j, \ldots, 2n + j\} & \text{if } j < 2n \\ \{j - 2n, \ldots, 6n + 1 - j\} & \text{if } j > 2n. \end{cases}$$

In particular, whenever the i, j entry of A_n is 1 (with j odd), so is m_{ij}. Similarly, for j even, I_j covers I_i for

$$i \in \begin{cases} \{2n - j, \ldots, 2n - 1 + j\} & \text{if } j \leq 2n \\ \{j - 2n + 1, \ldots, 6n + 2 - j\} & \text{if } j > 2n. \end{cases}$$

Thus the only entries of M which disagree with those of A_n are the pairs with

$$i = 2n + j \quad \text{if } j \leq 2n$$

and

$$i = 2n - j \quad \text{if } j \geq 2n + 2.$$

Since M is symmetric, we have (using *11.11(iii)*)

$$\exp h(\theta) = spec(M) = \|M\| = \|\hat{M}\|.$$

The kernel for $\hat{M}$ is the characteristic function of a set D which misses the diamond $\Diamond$ on $2n$ half-squares, each of side $\frac{1}{4n}$. (Note that D also contains some points outside $\Diamond$.) The measure of the union of these squares is

$$\frac{2n}{2} \left[\frac{1}{4n} \right]^2 = \frac{1}{16n}.$$

It follows that

$$\| \frac{1}{4n} \hat{M} \| \geq \|C\| - \frac{1}{4\sqrt{n}} = \frac{2}{\pi} - \frac{1}{4\sqrt{n}}. \quad ∎$$

Now we invoke *11.7, 11.13* and *11.17* to get

11.18. PROPOSITION. *If* $4n + 1 \leq m \leq 4n + 4$, *then*

$$4n \left[\frac{2}{\pi} - \frac{1}{4\sqrt{n}}\right] \leq \exp H(\mathfrak{C}_m) \leq 4(n + 1) \left[\frac{2}{\pi} + \frac{1}{2\sqrt{n + 1}}\right].$$

PROOF: By *11.7*,

$$\exp H(\mathfrak{C}_{4n+1}) \leq \exp H(\mathfrak{C}_m) \leq \exp H(\mathfrak{C}_{4n+4}).$$

On the left, *11.17* gives

$$\exp H(\mathfrak{C}_{4n+1}) \geq 4n \left[\frac{2}{\pi} - \frac{1}{4\sqrt{n}}\right].$$

On the right, we have by *11.10* and *11.13*

$$\exp H(\mathfrak{C}_{4n+4}) \leq spec(\hat{A}_{2n+2}) \leq 2(2n + 2) \left[spec(C) + \frac{1}{2(2n + 2)}\right]$$

$$= (4n + 4) \left[\frac{2}{\pi} + \frac{1}{2\sqrt{n + 1}}\right]. \blacksquare$$

As a corollary of *11.18*, we have, for $4n + 1 \leq m \leq 4n + 4$,

$$\frac{4n}{m} \left[\frac{2}{\pi} - \frac{1}{4\sqrt{n}}\right] \leq \frac{1}{m} \exp H(\mathfrak{C}_m) \leq \frac{4(n + 1)}{m} \left[\frac{2}{\pi} + \frac{1}{2\sqrt{n + 1}}\right].$$

Estimating the quantities on the left in terms of m, we have

$$\frac{4n}{m} \geq \frac{m - 4}{m} = 1 - \frac{4}{m}$$

and

$$\frac{1}{4\sqrt{n}} = \frac{1}{2\sqrt{4n}} \leq \frac{1}{2\sqrt{m - 4}}$$

while on the right,

$$\frac{4(n + 1)}{m} \leq \frac{m + 3}{m} = 1 + \frac{3}{m}$$

and

$$\frac{1}{2\sqrt{n + 1}} = \frac{1}{\sqrt{4n + 4}} \leq \frac{1}{\sqrt{m}}.$$

Thus, we have for all m the estimates

$$\left(1 - \frac{4}{m}\right) \left[\frac{2}{\pi} - \frac{1}{2\sqrt{m - 4}}\right] \leq \frac{1}{m} \exp H(\mathfrak{C}_m) \leq \left(1 + \frac{3}{m}\right) \left[\frac{2}{\pi} + \frac{1}{\sqrt{m}}\right].$$

But now, taking $m \to \infty$, we have that $\lim_{m\to\infty} \frac{1}{m} \exp H(\mathfrak{C}_m)$ exists and equals $\frac{2}{\pi}$. As a consequence, $\liminf_{m\to\infty} \frac{1}{m} \exp H(\mathfrak{S}_m) \geq \frac{2}{\pi}$; but since by *11.15* we have $\limsup_{m\to\infty} \frac{1}{m} \exp H(\mathfrak{S}_m) \leq \frac{2}{\pi}$, we have proved the desired equation in *11.6*:

$$\lim_{n\to\infty} \frac{1}{n} \exp H(\mathfrak{C}_n) = \lim_{n\to\infty} \frac{1}{n} \exp H(\mathfrak{S}_n) = \frac{2}{\pi}.$$

This completes the proof of *11.6*. $\blacksquare$

REFERENCES

[**ALM**] L. Alsedà, J. Llibre, & M. Misiurewicz, *Periodic orbits of maps of* Y, Trans. A.M.S. **313** (1989), 475-538.

[**ALS**] L. Alsedà, J. Llibre, & R. Serra, *Minimal periodic orbits for continuous maps of the interval*, Trans. A.M.S. **286** (1984), 595-627.

[**Ba**] S. Baldwin, *Generalizations of a theorem of Sarkovskii on orbits of continuous real valued functions*, Discrete Math **67** (1987), 111-127.

[**BCJM**] L. Block, E. Coven, L. Jonker, & M. Misiurewicz, *Primary cycles on the circle*, Trans. A.M.S. **311** (1989), 323-335.

[**Be1**] C. Bernhardt, *The ordering on permutations induced by continuous maps of the real line*, Ergod. Thy. & Dyn. Syst. **7** (1987), 155-160.

[**Be2**] _____, *Oriented Markov graphs of the interval*, preprint, Lafayette College, 1984 (unpublished).

[**Be3**] _____, *Simple permutations with order a power of two*, Ergod. Thy. & Dyn. Syst. **4** (1984), 179-186.

[**BGMY**] L. Block, J. Guckenheimer, M. Misiurewicz, & L.-S. Young, *Periodic points and topological entropy of one-dimensional maps*, Lect. Notes in Math. **819** Global Theory of Dynamical Systems (1980), 18-34.

[**Bl**] L. Block, *Simple periodic orbits of mappings of the interval*, Trans. A.M.S. **254** (1979), 391-398.

[**BlC**] L. Block & W. A. Coppel, *Stratification of continuous maps of an interval*, Trans. A.M.S. **297** (1986), 587-604.

[**BlCv1**] L. Block & E. Coven, *Topological conjugacy and transitivity for a class of piecewise monotone maps of the interval*, Trans. A.M.S. **300** (1987), 297-306.

[**BlCv2**] _____, *Approximating entropy of maps of the interval*, in "Proceedings of the Semester on Ergodic Theory and Dynamical Systems," Banach Center Publ. 23, Polish Acad. Sci., Warsaw, 1989, pp. 237-242.

[**BlH**] L. Block & D. Hart, *Orbit types for maps of the interval*, Ergod. Thy. & Dyn. Syst. **7** (1987), 161-164.

[**Bu**] U. Burkart, *Interval mapping graphs and periodic points of continuous functions*, J. Comb. Theory (**B**) **32** (1982), 57-68.

[**C**] W. A. Coppel, *Šarkovskii-minimal orbits*, Math. Proc. Camb. Phil. Soc. **93** (1983), 397-408.

[**Ca**] L. Carvalho, *On an extension of Sarkovskii's order*, J. Math. Anal. & Appl. **138** (1989), 52-58.

[**CE**] P. Collet & J.-P. Eckmann, "Iterated Maps on the Interval as Dynamical Systems," Birkhäuser, Basel, Boston, Stuttgart, 1980.

[**F0**] В. В. Федоренко [V. V. Fedorenko], Канонические периодические траектории одномерных динамических систем *[Canonical periodic trajectories of one-dimensional dynamical systems]*, in "Приближенные и Качественные Методы Дифференциально функциональных Уравнений [Approximate and qualitative methods in functional differential equations]," Инст. Мат. АН УССР [Inst. Math., Ukrainian Acad. Sci.], Kiev, 1983, pp. 106-109.

[**F1**] _____, Свойства циклических подстановок, используемые в одномерних динамических системах *[Properties of cyclic permutations, which are used in one-dimensional dynamical systems]*, in "Дифференциально-функциональные Уравнения и их Приложения [Functional-differential equations and their applications]," Инст. Мат. АН УССР [Inst. Math., Ukrainian Acad. Sci.], Kiev, 1985, pp. 80-86.

[**F2**] _____, Частично-упорядоченное множество типов периодических траекторий одномерныих динамических систем *[Partially ordered set of types of periodic trajectories of one-dimensional dynamical systems]*, in "Динамические Системы и Дифференциально-разностные Уравнения [Dynamical systems and differential-difference equations]," Инст. Мат. АН УССР [Inst. Math., Ukrainian Acad. Sci.], Kiev, 1986, pp. 90-97.

[**GT**] W. Geller & J. Tolosa, *Maximal entropy odd orbit types*, preprint, Univ. Calif., Berkeley 1989 (to appear, Trans. A.M.S.).

[**H1**] C.-W. Ho, *On Block's condition for simple periodic orbits of functions on an interval*, Trans. A.M.S. **281** (1984), 827-832.

[**H2**] _____, *On the structure of the minimum orbits of periodic points for maps of the real line*, preprint, Southern Ill. Univ. Edwardsville, 1982 (unpublished).

[**HM**] C.-W. Ho & C. Morris, *A graph theoretic proof of Sharkovsky's theorem on the periodic points of continuous functions*, Pacific J. Math. **96** (1981), 361-370.

109

[J] I. Jungreis, *Some results on the Sarkovskii partial ordering of permutations*, preprint, Harvard Univ., 1987 (to appear, Trans. A.M.S.).

[LMPY] T.-Y. Li, M. Misiurewicz, G. Pianigiani, & J. Yorke, *No division implies chaos*, Trans. A.M.S. **273** (1982), 191-199.

[M] M. Misiurewicz, *Horseshoes for mappings of the interval*, Bull. Acad. Polon. Sci. **27** (1979), 167-169.

[MS] M. Misiurewicz & W. Szlenk, *Entropy of piecewise monotone mappings*, Studia Math. **67** (1980), 45-63.

[MSS] N. Metropolis, M.L. Stein & P. R. Stein, *On finite limit sets for transformations on the unit interval*, J. Comb. Theory **(A)** 15 (1973), 25-44.

[MT] J. Milnor & W. Thurston, *On iterated maps of the interval*, Lect. Notes in Math. **1342** Dynamical Systems (1988), 465-563.

[My] P. J. Myrberg, *Iteration der Reellen Polynome zweiten Grades*, Annales Academiae Scientaiarum Fennicae **(A)** 336 (1963), 3-18.

[N] Z. Nitecki, *Topological dynamics on the interval*, in "Ergodic Theory and Dynamical Systems II (Proc. Special Year, Maryland 1979-80)," ed. A. Katok, Birkhäuser, Basel, Boston, & Stuttgart, 1982, pp. 1-73.

[S] E. Seneta, "Non-Negative Matrices: An Introduction to Theory and Applications (2 ed.)," Springer-Verlag, New York, Heidelberg, Berlin, 1981.

[Sa] А. Н. Шарковский [A. N. Sharkovskiĭ], Сосуществование циклов непрерывного преобразования прямой в себя *[Coexistence of cycles of a continuous map of the line into itself]*, Укр. Мат. Ж. [Ukrain. Math. J.] 16 (1964), 61-71.

[St] P. Štefan, *A theorem of Šarkovskii on the existence of periodic orbits of continuous endomorphisms of the real line*, Comm. Math. Physics **54** (1977), 237-248.

[Str] P. D. Straffin, *Periodic points of continuous functions*, Math. Mag. **51** (1978), 99-105.

[T] Y. Takahashi, *A formula for topological entropy of one-dimensional dynamics*, Sci. Papers College Gen. Ed. Univ. Tokyo **30** (1980), 11-22.

[W] H. Wielandt, *Unzerlegbare, nicht negative Matrizen*, Math. Zeitschrift. **52** (1950), 642-8.

INSTYTUT MATEMATYKI, UNIWERSYTET WARSZAWSKI, PAŁAC KULTURY I NAUKI, IX P., 00-901 WARSZAWA, POLAND

DEPARTMENT OF MATHEMATICS, TUFTS UNIVERSITY, MEDFORD, MA 02155, USA

MEMOIRS of the American Mathematical Society

SUBMISSION. This journal is designed particularly for long research papers (and groups of cognate papers) in pure and applied mathematics. The papers, in general, are longer than those in the TRANSACTIONS of the American Mathematical Society, with which it shares an editorial committee. Mathematical papers intended for publication in the Memoirs should be addressed to one of the editors:

Ordinary differential equations, partial differential equations and applied mathematics to ROGER D. NUSSBAUM, Department of Mathematics, Rutgers University, New Brunswick, NJ 08903

Harmonic analysis, representation theory and Lie theory to AVNER D. ASH, Department of Mathematics, The Ohio State University, 231 West 18th Avenue, Columbus, OH 43210

Abstract analysis to MASAMICHI TAKESAKI, Department of Mathematics, University of California, Los Angeles, CA 90024

Real and harmonic analysis to DAVID JERISON, Department of Mathematics, M.I.T., Rm 2–180, Cambridge, MA 02139

Algebra and algebraic geometry to JUDITH D. SALLY, Department of Mathematics, Northwestern University, Evanston, IL 60208

Geometric topology and general topology to JAMES W. CANNON, Department of Mathematics, Brigham Young University, Provo, UT 84602

Algebraic topology and differential topology to RALPH COHEN, Department of Mathematics, Stanford University, Stanford, CA 94305

Global analysis and differential geometry to JERRY L. KAZDAN, Department of Mathematics, University of Pennsylvania, E1, Philadelphia, PA 19104-6395

Probability and statistics to RICHARD DURRETT, Department of Mathematics, Cornell University, Ithaca, NY 14853-7901

Combinatorics and number theory to CARL POMERANCE, Department of Mathematics, University of Georgia, Athens, GA 30602

Logic, set theory, general topology and universal algebra to JAMES E. BAUMGARTNER, Department of Mathematics, Dartmouth College, Hanover, NH 03755

Algebraic number theory, analytic number theory and modular forms to AUDREY TERRAS, Department of Mathematics, University of California at San Diego, La Jolla, CA 92093

Complex analysis and nonlinear partial differential equations to SUN-YUNG A. CHANG, Department of Mathematics, University of California at Los Angeles, Los Angeles, CA 90024

All other communications to the editors should be addressed to the Managing Editor, DAVID J. SALTMAN, Department of Mathematics, University of Texas at Austin, Austin, TX 78713.

General instructions to authors for

PREPARING REPRODUCTION COPY FOR MEMOIRS

> For more detailed instructions send for AMS booklet, "A Guide for Authors of Memoirs."
> Write to Editorial Offices, American Mathematical Society, P.O. Box 6248,
> Providence, R.I. 02940.

MEMOIRS are printed by photo-offset from camera copy fully prepared by the author. This means that the finished book will look exactly like the copy submitted. Thus the author will want to use a good quality typewriter with a new, medium-inked black ribbon, and submit clean copy on the appropriate model paper.

Model Paper, provided at no cost by the AMS, is paper marked with blue lines that confine the copy to the appropriate size.

Special Characters may be filled in carefully freehand, using dense black ink, or **INSTANT** ("rub-on") **LETTERING** may be used. These may be available at a local art supply store.

Diagrams may be drawn in black ink either directly on the model sheet, or on a separate sheet and pasted with rubber cement into spaces left for them in the text. Ballpoint pen is not acceptable.

Page Headings (Running Heads) should be centered, in CAPITAL LETTERS (preferably), at the top of the page — just above the blue line and touching it.

LEFT-hand, EVEN-numbered pages should be headed with the AUTHOR'S NAME;

RIGHT-hand, ODD-numbered pages should be headed with the TITLE of the paper (in shortened form if necessary).

Exceptions: PAGE 1 and any other page that carries a display title require NO RUNNING HEADS.

Page Numbers should be at the top of the page, on the same line with the running heads.

LEFT-hand, EVEN numbers — flush with left margin;

RIGHT-hand, ODD numbers — flush with right margin.

Exceptions: PAGE 1 and any other page that carries a display title should have page number, centered below the text, on blue line provided.

FRONT MATTER PAGES should be numbered with Roman numerals (lower case), positioned below text in same manner as described above.

MEMOIRS FORMAT

> It is suggested that the material be arranged in pages as indicated below.
> Note: <u>Starred items (*)</u> are requirements of publication.

Front Matter (first pages in book, preceding main body of text).

Page i — *Title, *Author's name.

Page iii — Table of contents.

Page iv — *Abstract (at least 1 sentence and at most 300 words).

> Key words and phrases, if desired. (A list which covers the content of the paper adequately enough to be useful for an information retrieval system.)

> *<u>1991 Mathematics Subject Classification</u>. This classification represents the primary and secondary subjects of the paper, and the scheme can be found in Annual Subject Indexes of MATHEMATICAL REVIEWS beginnning in 1990.

Page 1 — Preface, introduction, or any other matter not belonging in body of text.

> Footnotes: *Received by the editor date.
> Support information — grants, credits, etc.

First Page Following Introduction – Chapter Title (dropped 1 inch from top line, and centered). Beginning of Text.

Last Page (at bottom) – Author's affiliation.